AF658780

FANS AND COMPRESSORS HANDBOOK

FANS AND COMPRESSORS HANDBOOK

Patrick Rebello

Oxford Book Company
Jaipur ▪ New Delhi

Published By

Oxford Book Company

Regd. Off. "Bony Residency", Gate No. 2
Opp. Tilak Public School, Vishweswariya Nagar,
Gopalpura Road, Jaipur - 302018, India
Ph: 0141 - 2761280, Telefax: 2761381
email - oxfordbookjpr@gmail.com
website - oxfordbookcompany.com

Delhi Office: 102, 1st Floor, Satyam House,
4327/3, Ansari Road, Darya Ganj,
New Delhi - 110002, India
Ph: 011-45652440

Distributed By
KEYA INTERNATIONAL PRESS LLP
E-mail: keyainternationalpressllp@gmail.com

Published in India, 2025

© All Rights Reserved

ISBN : 9789391096731

All Rights Reserved. No part of this publication may be reproduced, stored in a retrieval system, or transmitted, in any form or by any means, electronic, mechanical, photocopying, recording, scanning or otherwise, without the prior written permission of the copyright owner.

Typeset by Shivangi Computers, Jaipur

Printed and bound in India by Thomson Press Ltd.,
New Delhi

Contents

Preface

Fans and blowers provide air for ventilation and industrial process requirements. Fans generate a pressure to move air (or gases) against a resistance caused by ducts, dampers, or other components in a fan system. The fan rotor receives energy from a rotating shaft and transmits it to the air. Fans, blowers and compressors are differentiated by the method used to move the air, and by the system pressure they must operate against. As per American Society of Mechanical Engineers (ASME) the specific ratio - the ratio of the discharge pressure over the suction pressure - is used for defining the fans, blowers and compressors. Fan and blower selection depends on the volume flow rate, pressure, type of material handled, space limitations, and efficiency. Fan efficiencies differ from design to design and also by types. Fan characteristics can be represented in form of fan curve(s). The fan curve is a performance curve for the particular fan under a specific set of conditions. The fan curve is a graphical representation of a number of inter-related parameters. Typically a curve will be developed for a given set of conditions usually including: fan volume, system static pressure, fan speed, and brake horsepower required to drive the fan under the stated conditions. Some fan curves will also include an efficiency curve so that a system designer will know where on that curve the fan will be operating under the chosen conditions. In the many curves shown in the Figure, the curve static pressure (SP) vs. flow is especially important.

The intersection of the system curve and the static pressure curve defines the operating point. When the system resistance changes, the operating point also changes. Once the operating point is fixed, the power required could be found by following a vertical line that passes through the operating point to an intersection with the power (BHP) curve. A horizontal line drawn through the intersection with the power curve will lead to the required power on the right vertical axis. In the depicted curves, the fan efficiency curve is also presented. Precise determination of air-flow and required outlet pressure are most important in proper selection of fan type and size. The air-flow required depends on the process requirements; normally determined from heat transfer rates, or combustion air or flue gas quantity to be handled. System pressure requirement is usually more difficult to compute or predict. Detailed analysis should be carried out to determine pressure drop across the length, bends, contractions and expansions in the ducting system, pressure drop across ilters, drop in branch lines, etc. These pressure drops should be added to any fixed pressure required by the process (in the case of ventilation fans there is no fixed pressure requirement). Frequently, a very conservative approach is adopted allocating large safety margins, resulting in over-sized fans which operate at flow rates much below their design values and, consequently, at very poor efficiency. Once the system flow and pressure requirements are determined, the fan and impeller type are then selected. For best results, values should be obtained from the manufacturer for specific fans and impellers. The choice of fan type for a given application depends on the magnitudes of required flow and static pressure. For a given fan type, the selection of the appropriate impeller depends additionally on rotational speed. Speed of operation varies with the application. High speed small units are generally more economical because of their higher hydraulic efficiency and relatively low cost. However, at low pressure ratios, large, low-speed units are preferable. In the case of centrifugal fans, the hub-to-tip ratios (ratio of inner-toouter impeller diameter) the tip angles (angle at which forward or backward curved blades are curved at the blade tip - at the base the blades are always oriented in the direction of flow), and the blade width determine the pressure developed by the fan. Forward curved fans have large hub-to-tip ratios compared to backward curved fans and produce lower pressure. Radial fans can be made with different heel-to-tip ratios to produce different pressures.

This Handbook will give readers, and student community in particular a vivid picture of Fan and compressors technology.

— Patrick Rebello

Chapter 1

Types of Compressors

The three basic types of air compressors are

- reciprocating
- rotary screw
- rotary centrifugal.

These types are further specified by:

- the number of compression stages
- cooling method (air, water, oil)
- drive method (motor, engine, steam, other)
- lubrication (oil, Oil-Free where Oil Free means no lubricating oil contacts the compressed air)
- packaged or custom-built.

Reciprocating Air Compressors

Reciprocating air compressors are positive displacement machines, meaning that they increase the pressure of the air by reducing its volume. This means they are taking in successive volumes of air which is confined within a closed space and elevating this air to a higher pressure. The reciprocating air compressor accomplishes this by a piston within a cylinder as the compressing and displacing element.

Single-stage and two-stage reciprocating compressors are commercially available.

- Single-stage compressors are generally used for pressures in the range of *70 psig to 100 psig.*

Two-stage compressors are generally used for higher pressures in the range of *100 psig to 250 psig.*

Note that

- *1 HP ~ 4 CFM at 100 psi*

and that *1 to 50 HP* are typically for reciprocating units. Compressors *100 hp* and above are typically Rotary Screw or Centrifugal Compressors.

The reciprocating air compressor is single acting when the compressing is accomplished using only one side of the piston. A compressor using both sides of the piston is considered double acting.

Load reduction is achieved by unloading individual cylinders. Typically this is accomplished by throttling the suction pressure to the cylinder or bypassing air either within or outside the compressor. Capacity control is achieved by varying speed in engine-driven units through fuel flow control. Reciprocating air compressors are available either as air-cooled or water-cooled in lubricated and non-lubricated configurations and provide a wide range of pressure and capacity selections.

Rotary Screw Compressors

Rotary air compressors are positive displacement compressors. The most common rotary air compressor is the single stage helical or spiral lobe oil flooded screw air compressor. These compressors consist of two rotors within a casing where the rotors compress the air internally. There are no valves. These units are basically oil cooled (with air cooled or water cooled oil coolers) where the oil seals the internal clearances. Since the cooling takes place right inside the compressor, the working parts never experience extreme operating temperatures. The rotary compressor, therefore, is a continuous duty, air cooled or water cooled compressor package.

Rotary screw air compressors are easy to maintain and operate. Capacity control for these compressors is accomplished by variable speed and variable compressor displacement. For the latter control technique, a slide valve is positioned in the casing. As the compressor capacity is reduced, the slide valve opens, bypassing a portion of the compressed air back to the suction. Advantages of the rotary screw compressor include smooth, pulse-free air output in a compact size with high output volume over a long life.

The oil free rotary screw air compressor utilises specially designed air ends to compress air without oil in the compression chamber yielding true oil free air. Oil free rotary screw air compressors are available air cooled and water cooled and provide the same flexibility as oil flooded rotaries when oil free air is required.

Centrifugal Compressors

The centrifugal air compressor is a dynamic compressor which depends on transfer of energy from a rotating impeller to the air.

Centrifugal compressors produce high-pressure discharge by converting angular momentum imparted by the rotating impeller (dynamic displacement). In order to do this efficiently, centrifugal compressors rotate at higher speeds than the other types of compressors. These types of compressors are also designed for higher capacity because flow through the compressor is continuous.

Adjusting the inlet guide vanes is the most common method to control capacity of a centrifugal compressor. By closing the guide vanes, volumetric flows and capacity are reduced. The centrifugal air compressor is an oil free compressor by design. The oil lubricated running gear is separated from the air by shaft seals and atmospheric vents. The pressure loss through air compressor inlets should be kept to a minimum to avoid noise problems. The diagram below indicates reasonable maximum flow through inlet pipes.

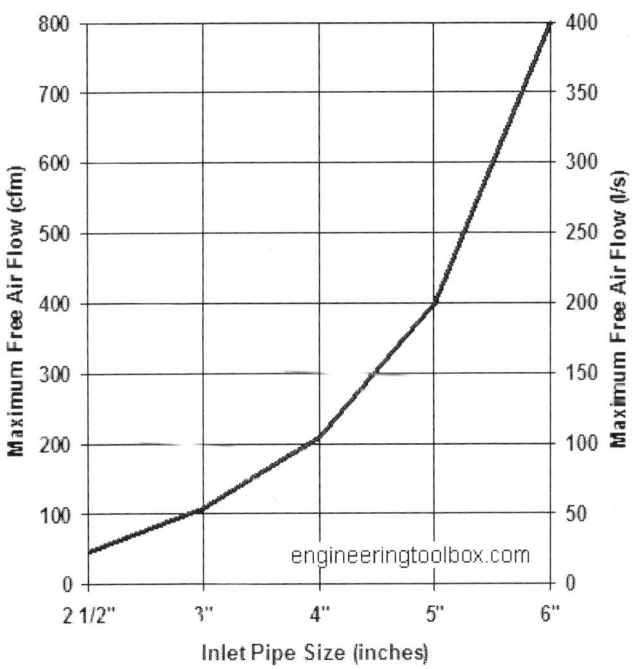

- *1 cfm = 0.47 l/s*
- *1 inch = 25.4 mm*

About the ToolBox

We appreciate any comments and tips on how to make The Engineering ToolBox a better information source. Please contact us by email if You find any faults, inaccuracies, or otherwise unacceptable information.

The content in The Engineering ToolBox is copyrighted but can be used with NO WARRANTY or LIABILITY. Important information should always be double checked with alternative sources. All applicable national and local regulations and practices concerning this aspects must be strictly followed and adhered to.

Micro-Design, Inc. is a Texas Corporation offering products and services in multiple industries:

- CNG Compressed Natural Gas (CNG)
  - Compressors, Compressor Packages
  - Valve Panels, Controls, Local SCADA
  - CNG Documents
- Oil & Gas
  - Natural Gas Compressor Packages
  - Compressor Controls
  - Complete Oil Well Site Monitoring and Control
  - Remote Webcam
- Water & Waste Water
  - Plant instrumentation and controls
  - SCADA system, programming
  - Pump Station Controls
  - Remote Monitoring
- Asset Management
  - LevelCon Web based Tank Monitoring – remote above ground or underground
  - Local and Remote Stationary Tank Monitors with Overfill Protection, Automatic Tank Gauging Systems (ATG)
  - Mobile Asset Tracking – vehicles, trucks, containers, backup generators, construction equipment, mobile road signs, fleets
  - Web based Scada Integration – remote instruments, flow meters, pumps, power usage,

- Security Systems
    - o Site Security – Local Ethernet based LevelCon monitoring
    - o Web based, wireless, solar powered security Cams
    - o Construction site Security
    - o Event based Site security.

Micro-Design, Inc. has been in the CNG industry since 1991. Originally the company provided state-of-the-art controls for operating CNG stations. Very cost-effective and efficient solutions have been developed based on 15 years prior experience in the Oil & Gas industry. MDi's knowledge base and experience in CNG increased considerably with the deployment of over 300 CNG stations controls worldwide. Subsequently, MDi brought the proper resources into the company to provide complete CNG station solutions including design, manufacture, installation and ongoing support. The initial CNG compressor skids, dispensers, valve panels, controls, etc. were deployed in 1999.

Today, MDi is supporting several complete CNG stations deployed by MDi worldwide and also individual components such as compressors, valve panels, dispensers and controls purchased by other CNG solution providers. We are certified distributors of GE Gemini compressors and spare parts. We can supply GE compressors as well as MDi's own exclusive CNG compressor design manufactured by CIP Compressors, Inc. MDi is also offering a unique design for low inlet pressure applications which involves a Screw compressor on the front end coupled with a reciprocating compressor to handle the high pressure side. Our controls fit any cng application, they are easy to use and incorporate MDi proprietary LevelCon web based remote monitoring system They can also be easily incorporated into any SCADA system.

MDi is ready to assist in new CNG station designs, sizing, equipment supply, installation, startup and commissioning. We pride ourselves in customer support and log term relationships.

Axial-flow Compressors: A Strategy for Aerodynamic Design and Analysis

Modern axial-flow compressor aerodynamic design and analysis technology is so often proprietary, and the role of computers is so pervasive that about the only way this subject can be described today is to concentrate on a specific system. The author provides a complete description of an actual working system, with particular emphasis on:

- Practical methodology that can easily be implemented on a personal computer
- Detailed descriptions sufficient to implement and apply all or part of the system
- Integration of the various methods into an efficient design and analysis system
- Core topics of performance prediction, aerodynamic design, and internal flow analysis
- Novel topics, such as an efficient multi-stage compressor design procedure and application to working fluids requiring a non-ideal gas equation of state
- Many design examples to evaluate and demonstrate a variety of design strategies.

Mulistage Centrifugal Compressor and Fan Design

- Mixed axial/radial flow path design "from scratch" using a small number of basic parameters and a set of geometric constraints.
- Corporate-value features: loadable custom-defined fluid models, loss models and material databases.
- Camberline/Thickness interactive profiling.
- Meanline analysis of multistage mixed axial/radial compressor flow path using as-designed and off-design operational conditions.
- Integrated multi-variant optimiser capable of developing centrifugal compressor maps and searching for optimal solution.
- Performance map generation for more effective fan and centrifugal compressor design.
- Capability to export 3D compressor airfoil geometry in formats acceptable for Numeca, CFX, etc. CFD solvers and in IGES format for CAD/CAE packages.
- Express 3D structural and modal analysis (including Campbell diagram) of centrifugal compressor blade via embedded FEA module with automatic mesh generation – AxSTRESS.
- 3D CFD flow analysis via embedded CFD calculation solver AxCFD with automatic mesh generation and post-processing of obtained results.

- Flow charts and charts for blade loading on Profiling module.
- >Detailed Mollier diagram.
- 1D/2D Streamline solver allows finding compressor surge, stall and choke. It is capable to calculate transonic and subsonic stages of compressor.
- AxMAP helps develop compressor map and can show dependence of pressure ratio of mass flow and rotation speed.

Fast Turbocharger Design

TurboMatch software from Concepts NREC (conceptsnrec.com) bridges the gap between engines and turbochargers by allowing users to design and analyse turbochargers optimally matched to a specific engine system. The software consists of two modes: Design, wherein users can size the compressor and turbine to the required pressure ratios to match the power and rotational speed, and Analysis, where users can manipulate geometries and examine system performance as it interacts with the compressor and turbine maps. TurboMatch permits the design of turbines with waste gates, variable geometry turbines, intercoolers, and high- and low-pressure exhaust gas recirculation (EGR) systems. The software replaces the trial-and-error design methods, condensing the design

Energy efficiency benefits of variable speed control in pumps, fans and Compressors

Valves, taps and gates are the most commonly used devices. However, substantial energy savings can be obtained by using variable speed drives to control the flow rate or pressure in pumps, fans and compressors. In pumping applications, the most significant savings are achieved with centrifugal pumps. This paper describes how centrifugal pumps, fans and compressors function in different operating modes and quantifies savings that speed control generates. Other advantages of this technique in terms of energy efficiency are also reviewed.

Fans and Compressors and for Chemical Applications

The CU230P-2 benefits from a simplified engineering via an integrated USB interface and can be operated with the new IOP (Intelligent Operator Panel). The integrated MMC card and so-called clone functionality enable data to be transferred easily from one drive to another. The new Control Unit permits the direct control of shutoff valves without the need for any extra external components. Furthermore, it is designed for the connection of an extensive range of temperature sensors.

Sinamics G120 inverters can now be extended by an additional module: namely, the Control Unit CU240S DP-F that is specially made for applications in the chemical industry. This module meets the NAMUR standards in every respect and includes an ATEX-certified PTC evaluation function.

This new Control Unit is an integral part of the modular-design Sinamics inverter series and can be used in combination with all available Power Modules, including those variants with energy-saving regenerative feedback capability. Equipped with the new PM240 Power Module in frame size GX, the Sinamics G120 inverters are now available with a power output of 250 kW. As the new variant of Power Module with regenerative feedback capability, the PM250 without sine-wave filter also makes it possible to realise cost-effective solutions with integrated energy recovery in regenerative mode. This variant is designed for the power range from 18.5 kW to 90 kW.

About Siemens

The Siemens Industry Sector (Erlangen, Germany) is the worldwide leading supplier of production, transportation, building and lighting technologies. With integrated automation technologies as well as comprehensive industry-specific solutions, Siemens increases the productivity, efficiency and flexibility of its customers in the fields of industry and infrastructure. The Sector consists of six Divisions: Building Technologies, Drive Technologies, Industry Automation, Industry Solutions, Mobility and Osram. With around 222,000 employees worldwide Siemens Industry achieved in fiscal 2008 a profit of EUR3.86 billion with revenues totalling EUR38 billion.

About Siemens Drive Technologies Division

The Siemens Drive Technologies Division (Nuremberg, Germany) is the world s leading supplier of products and services for production machinery and machine tools. Drive Technologies offers integrated technologies that cover the entire drive train with electrical and mechanical components. This includes standard products but also encompasses industry-specific control and drive solutions for metal forming, printing and electronic manufacturing as well as solutions for glass, wood, plastic, ceramic, textile and packaging equipment and crane systems. The services provided by the Division include mechatronics support in addition to online services for web-based fault management and preventive maintenance. With around 39,900 employees worldwide Siemens Drive Technologies achieved in fiscal 2008 total sales of EUR8.9 billion.

Design of Highly Loaded Axial Flow Fans and Compressors

In this monograph the author summarises the most important points he has learned during a career devoted to advanced fan and compressor research and development. Although all of this work was originally aimed at military aircraft engine applications, much of it is relevant to commercial aircraft engines and also to industrial and even consumer-oriented turbomachinery.

Guidance is given to the aerodynamic designer in most aspects of the detail design of axial compressors, backed up with references which the reader can pursue for a more comprehensive treatment of individual points. Practical and convenient design approaches are recommended wherever possible.

The author believes that the simplest approach which can be logically defended is usually the best one unless a designer has a very good reason for escalating the level of complexity in any area.

Fans, Compressors – the Right Drive Solution for Every Delivery Profile

There is no single ideal drive concept when it comes to energy efficiency and cost effectiveness in applications involving pumps, fans and compressors. Depending on customer requirements, four fundamentally different concepts are possible, for which Siemens provides energy-efficient drive solutions.

Two-point Control

Two-point control with soft starters is especially suitable for variable tapping such as level control in a tank. Thanks to on/off mode using the Sirius switching devices from Siemens, this drive concept ensures that sufficient liquid is always available in the case of variable tapping. The level is monitored by a sensor. If the level drops to the minimum, a pump switches on with a soft starter that then switches off again when the upper level is reached. The pump and tank are selected here such that there are relatively long on and off phases, and frequent switching operations are avoided. The energy consumption of this switching device drops here almost to that of a simple contactor. Thanks to its electronic generalised phase control at startup and rundown, the soft starter avoids severe shocks in the piping system and reduces wear and tear on the mechanical components. The motor and pump, for their part, run at rated speed with optimal efficiency and thus ensure efficient operation.

However, pumps, fans and compressors are often subject to stronger fluctuations in delivery rates. Two drive concepts are possible here with which energy requirements in the case of strong fluctuations in the delivery rate can be halved compared to a throttle control, for example, and in extreme cases, they can be reduced by up to 70 percent.

Soft Starter Frequency Inverter Cascade Circuit

In the case of the cascade circuit, the drive solution consists of one or more soft starters combined with a frequency inverter that handles fine adjustment. With a cascade circuit comprising a soft starter and a frequency inverter, the soft starter is automatically connected at a fixed speed after the maximum pumping capacity of the speed-controlled first pump has been exceeded. The frequency inverter now controls the precise requirement between 50 and 100 percent. On the one hand, this allows optimal utilisation of the advantages of both drive technologies frequency inverter and soft starter. On the other hand, by dividing the pumping capacity between two smaller units, delivery is significantly more energy-efficient, particularly at lower loads.

Cascade Circuit with Frequency Inverter Control Alone

If the process requires high control accuracy and speed, the obvious choice is to use frequency inverter control exclusively. Even with intense fluctuations in delivery rates, the frequency inverters adjust the motor speed quickly and precisely to the currently required delivery rate. Only the currently required energy is taken up, with maximum ease of control. With regard to avail- ability, the cascade circuit offers a further benefit thanks to redundant configuration of the pumps. No matter which of the four concepts is used when considering economic and ecological aspects: The components and systems of the product families Sirius switching devices, Sinamics frequency inverters, and Simotics motors can also be combined in individual ways. In this way, they enable optimal interaction with the relevant system components and seamless integration into the higher-level automation landscape.

The Industry Sector and Drive Technologies Division in Brief

The Siemens Industry Sector (Erlangen, Germany) is the world's leading supplier of innovative and environmentally friendly products and solutions for industrial customers. With end-to-end automation technology and industrial software, solid vertical-market expertise, and technology-based services, the Sector enhances its customers' productivity, efficiency, and flexibility. With a global workforce of more than 100,000 employees, the Industry Sector comprises the Divisions Industry Automation, Drive Technologies and Customer Services as well as the Business Unit Metals Technologies.

The Siemens Drive Technologies Division (Nuremberg, Germany) is the world's leading supplier of products, systems, applications, solutions and services for the entire drive train, with electrical and mechanical components. Drive Technologies serves all vertical markets in the production and process industries as well as the infrastructure/ energy segment. With its products and solutions, the Division fulfills the key requirements of its customers for productivity, energy efficiency, and reliability.

With pumps, fans and compressors, the right drive solution depends on the delivery profile in each case, and on the control accuracy required by an application. Deciding on the most suitable drive concept with regard to energy efficiency and cost effectiveness first requires an analysis of the application. This must take account not only of investment costs but also of plant-specific key data as well as the relevant delivery rate.

Throttle Control for the Upper Control Range

Throttle control represents the simplest and lowest- cost version to procure. It consists of a combination of circuit breaker and contactor in continuous operation, with the delivery rate controlled by an electromechanical throttle valve.

Since the motor operates throughout at rated speed, an energy-efficient motor such as the Simotics low- voltage asynchronous motor with IE3 efficiency is especially useful. However, this drive concept only pays off in the longer term when the delivery rate approximates to the maximum power and is subject to only slight or brief fluctuations.

Fluctuating delivery rates result in pronounced partial load losses, similar to driving a car at full acceleration and using the brake to control the speed.

Fans and Compressors

Now available from HMK Automation & Drives is the new Siemens Control Unit CU230P-2 for the Sinamics G120 frequency inverter series which is especially designed for use in applications using pumps, fans and compressors. The software includes a digital real-time clock, a hibernation, a flying restart and a PID controller module. The I/O and fieldbus interfaces such as Modbus RTU, CANopen, USS and Profibus have been fully adapted to the requirements of these applications. The CU230P-2 benefits from simplified engineering via an integrated USB interface and can be operated with the new IOP (Intelligent Operator Panel). The integrated MMC card and so-called clone functionality enable data to be transferred easily from one drive to another. The new Control Unit permits the direct control to shut off valves without the need for any extra external components.

The Control Unit is designed to facilitate the connection of an extensive range of temperature sensors. Sinamics G120 inverters can now be extended by an additional module: namely, the Control Unit CU240S DP-F which is specially made for applications in the chemical industry.

This module meets the NAMUR standards in every respect and includes an ATEX-certified PTC evaluation function. This new Control Unit is an integral part of the modular-design Sinamics inverter series and can be used in combination with all available Power Modules, including those variants with energy-saving regenerative feedback capability. Equipped with the new PM240 Power Module in frame size GX, the Sinamics G120 inverters are now available with a power output of 250 kW. The new variant of Power Module has regenerative feedback capabilities, the PM250 without sine-wave filter also makes it possible to realise cost-effective solutions with integrated energy recovery in regenerative mode. This variant is designed for a power range from 18.5 kW to 90 kW.

Design Considerations for the Volutes of Centrifugal Fans and Compressors

The initial conceptual design of centrifugal fan and compressor volutes is considered and extended to accommodate overhung volute designs often used in process and turbocharger compressors. The initial passage design is then developed through the application of a commercial computational fluid dynamics (CFD) code.' Based on the experimental data of a turbocharger compressor volute, three-dimensional, compressible, steady flow computations were carried out for alternative volute designs. Detailed internal flow data in both a conventional and a modified volute design, at both design and off-design flow conditions, are presented. The design investigation showed that enlarging the flow passage area near the tongue region, but without changing the exit-inlet area ratio of the volute, led to an improvement in the internal flow distribution at off-design flow conditions.

Compressor Map

Each compressor (or fan) in a gas turbine engine has an operating map. Complete maps are either based on compressor rig test results or are predicted by a special computer program. Alternatively the map of a similar compressor can be suitably scaled.

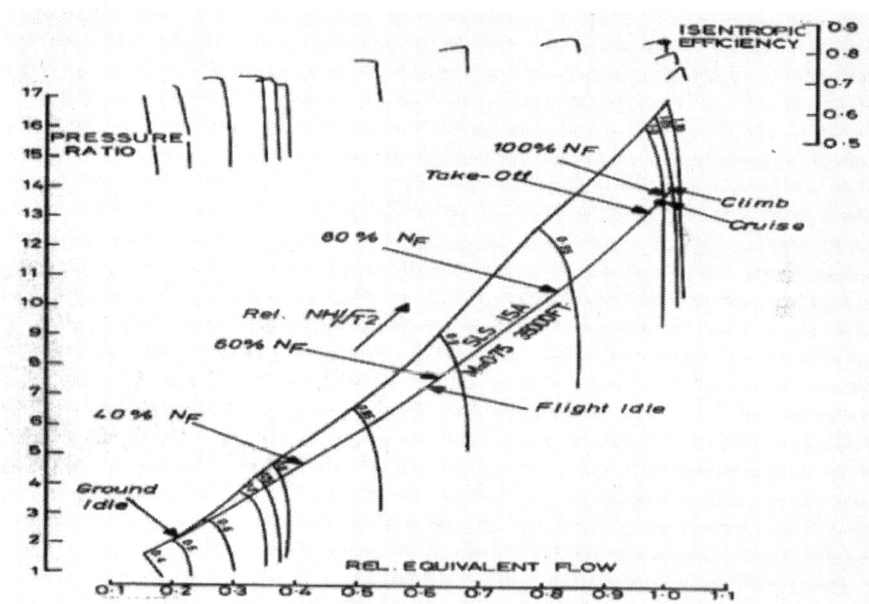

Figure: Typical high pressure compressor map

Compressor maps are important, since they are an integral part of predicting the performance of a gas turbine engine, both at design and off-design conditions. Fans and turbines also have operating maps, although the latter are significantly different in appearance to that of fans and compressors.

Flow Axis

The x-axis is usually some function of compressor entry mass flow, usually corrected flow or non-dimensional flow, as opposed to real flow. This axis can be considered a rough measure of the axial Mach number of the flow through the device.

Pressure Ratio Axis

Normally the y-axis is pressure ratio (P_{exit}/P_{inlet}), where P is stagnation (or total head) pressure.

$\Delta T/T$ (or similar), where T is stagnation (or total head) temperature, is also used.

Surge Line

The slightly kinked diagonal line on the main (i.e. lower) part of the map is known as the surge (or stall) line. Above this line is a region of unstable flow, which is an area best avoided.

A compressor surge, typically, causes an abrupt reversal of the airflow through the unit, as the pumping action of the aerofoils stalls (akin to an aircraft wing stalling).

Surge Margin

As the name suggests, surge margin provides a measure of how close an operating point is to surge. Unfortunately, there are a number of different definitions of surge margin. A popular one in use is defined as follows:

$$SM = 100\% \cdot \frac{\dot{m}_w - \dot{m}_s}{\dot{m}_w}$$

where:

$\dot{m}_w$ is the mass flow at the operating point, be it steady state or transient

$\dot{m}_s$ is the mass flow at surge, at same corrected speed as $\dot{m}_w$

Speed Lines

The slightly curved, near vertical, lines on the main part of the map are the (constant rotational) corrected speed lines. They are a measure of rotor blade tip Mach number. Note on the illustration that the speed lines are not distributed linearly with flow. This is because this particular compressor is fitted with variable stators, which open progressively as speed increases, causing an exaggerated increase in flow in the medium to high speed region. At low speed, the variable stators are locked, causing a more linear relationship between speed and flow.

Also note that beyond 100% flow, the speed lines close up rapidly, due to choking. Beyond choke, any further increase in speed will generate no further increase in airflow.

Efficiency axis

A sub-plot shows the variation of isentropic (i.e. adiabatic) efficiency with flow, at constant speed. Some maps use polytropic efficiency. Alternatively, for illustrative purposes, efficiency contours are sometimes cross-plotted onto the main map.

Note that the locus of peak efficiency exhibits a slight kink in its upward trend. This due to the choking-up of the compressor as speed increases, with the variable stators closed-off. The trend line resumes, once the variables start to move open.

Working Line

Also shown on the map, is a typical steady state working (or operating/running) line. This is a locus of the operating points of the engine, as it is throttled.

Being a high pressure ratio device, the working line is relatively shallow. If the unit had no variable geometry, there would be handling problems, because the surge line would be very steep and cross the working line at part-flow..

During a slam-acceleration from, say, a mid-throttle setting, the compressor working line will move rapidly towards surge and then slowly approach the steady state operating point, further up the map. The reverse effect occurs during a slam-deceleration. These effects are caused by the sluggish response of the spool (i.e. inertia effects) to rapid changes in engine fuel flow. Compressor surge is a particular problem during slamaccelerations and can be overcome by suitable adjustments to the fuelling schedule and/or use of blow-off (bleeding air off the compressor, for handling purposes). In the particular example shown, a slamaccel from ground idle would cause an HP compressor surge. Opening the blow-off would help, but some changes to the variable stator schedule might also be required. Because an HP compressor 'sees' the choked flow capacity of the HP turbine, the compressor working line is hardly affected by flight conditions. The slope of the working line approximates to a constant corrected outlet flow.

Fan Map

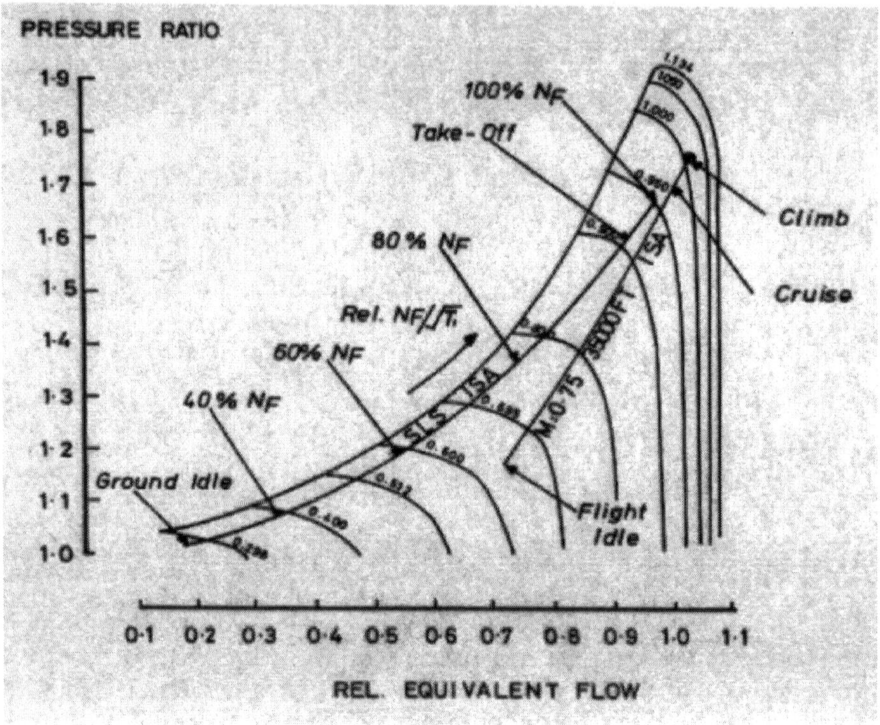

Figure: Typical high bypass ratio fan map

As the second illustration shows, a low pressure ratio fan (such as that used on a high bypass ratio turbofan) has a range of working lines. At high flight speeds, the ram pressure ratio factors up the cold nozzle pressure ratio, causing the nozzle to choke. Above the choking condition, the working lines tend to coalesce into a unique steep straight line. When the nozzle unchokes, the working line starts to become more curved, reflecting the curvature of the nozzle characteristic. With falling flight Mach number, the cold nozzle pressure ratio decreases. Initially this has no effect upon the position of the working line, apart from the curved (unchoked) tail, which becomes longer. Eventually, the cold nozzle will become unchoked at lower flight Mach numbers, even at full throttle. The working lines will now become curved, gradually migrating towards surge as flight Mach number decreases. The lowest surge margin working line occurs at static conditions.

Owing to the nature of the constraints involved, the fan working lines of a mixed turbofan are somewhat steeper than those of the equivalent unmixed engine.

The fan map shown is for the bypass (i.e. outer) section of the unit. The corresponding inner section map typically has longer, flatter, speed lines.

Military turbofans tend to have a much higher design fan pressure ratio than civil engines. Consequently the final (mixed) nozzle is choked at all flight speeds, over most of the throttle range. However, at low throttle settings the nozzle will unchoke, causing the lower end of the working lines to have a short curved tail, particularly at low flight speeds.

However, ultra-high bypass ratio turbofans have a very low design fan pressure ratio (e.g. 1.2, on the bypass section). Consequently, even at cruise flight speeds, the cold (or mixed final) propelling nozzle is unchoked. The fan working lines become more spread-out with flight Mach number, because they move bodily up the map, towards the top right hand corner. As a result, the static working line can be well into surge.

One solution is to have a variable area cold (or mixed) nozzle. Increasing the nozzle area at low flight speeds brings the fan working line away from surge.

An alternative solution is to fit a variable pitch fan. Scheduling the pitch of the fan blades has no impact upon the position of the fan working lines, but can be used to move the surge line upwards, to improve fan surge margin.

IP compressor map

Some turbofans have an Intermediate Pressure (IP) Compressor located between the fan and the HP compressor to increase overall pressure ratio. US civil engines tend to mount the IP compressor on the LP shaft, directly behind the fan, whereas Rolls-Royce normally mount the IP compressor on a separate (i.e. IP) shaft, which is driven by an IP turbine. Either way, matching problems can arise.

Basically the IP compressor outlet corrected flow must match the entry corrected flow of the HP compressor, which is decreasing as the engine is throttled back. Now, at a certain IP compressor working line slope, the IPC outlet corrected flow remains constant. However, by adopting a shallower working line, the extra IPC pressure ratio at a given IPC entry corrected flow enables the IPC outlet corrected flow to decrease and match up with the falling HP compressor entry corrected flow. Unfortunately this can lead to a poor IPC surge margin at part flow

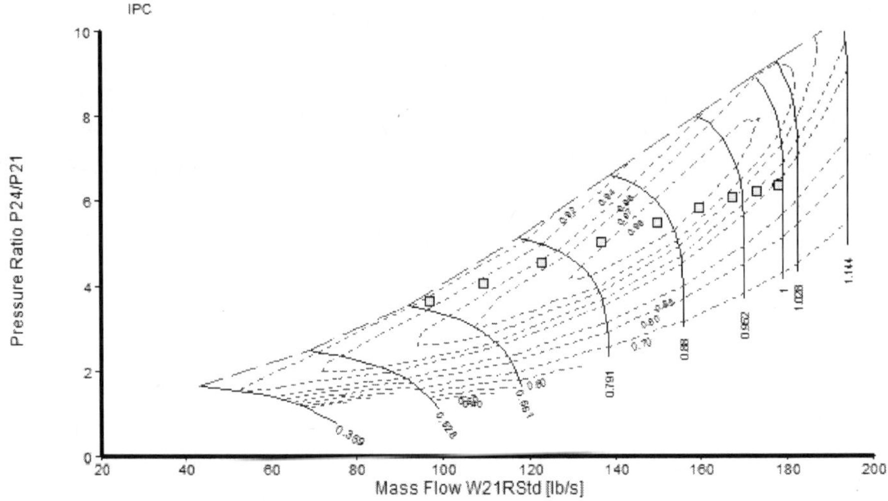

Figure: IPC working line, with blow-off valve closed throughout throttle range

At a given IPC pressure ratio, opening the blow-off valve forces the IP compressor entry corrected flow to increase, to a point where the IPC surge margin tends to be better. Effectively, opening the blow-off valve lowers the IPC working line. Any flow surplus to that demanded by the HP compressor passes through the blow-off valve into the bypass duct. The blow-off valve is normally only opened at throttled conditions, since it wastes energy.

Chapter 2

Air Compressor

An air compressor is a device that converts power (usually from an electric or diesel or gasoline engine) into kinetic energy by compressing and pressurising air, which, on command, can be released in quick bursts. There are numerous methods of air compression, divided into either positive-displacement or negative-displacement types

Positive-displacement air compressors work by forcing air into a chamber whose volume is reduced to effect the compression. Piston-type air compressors use this principle by pumping air into an air chamber through the use of the constant motion of pistons. They use unidirectional valves to guide air into a chamber, where the air is compressed Rotary screw compressors also use positive-displacement compression by matching two helical screws that, when turned, guide air into a chamber, the volume of which is reduced as the screws turn. Vane compressors use a slotted rotor with varied blade placement to guide air into a chamber and compress the volume.

Negative-displacement air compressors include centrifugal compressors. These devices use centrifugal force generated by a spinning impeller to accelerate and then decelerate captured air, which pressurises it Conventional air compressors are used in several different applications:

- To supply high-pressure clean air to fill gas cylinders
- To supply moderate-pressure clean air to a submerged surface supplied diver
- To supply moderate-pressure clean air for driving some office and school building pneumatic HVAC control system valves
- To supply a large amount of moderate-pressure air to power pneumatic tools

- For filling tires
- To produce large volumes of moderate-pressure air for macroscopic industrial processes (such as oxidation for petroleum coking or cement plant bag house purge systems).

Most air compressors are either reciprocating piston type or rotary vane or rotary screw. Centrifugal compressors are common in very large applications. There are two main types of air compressor's pumps: Oil lubed and oilless. The oilless system has more technical development, but they are more expensive, louder and last for less time than the oiled lube pumps. However, the air delivered has better quality.

Air Cycle Machine

An air cycle machine (ACM) is the refrigeration unit of the environmental control system (ECS) used in pressurised gas turbine-powered aircraft. Normally an aircraft has two or three of these ACM. Each ACM and its components are often referred as an Air Conditioning Pack. The air cycle cooling process uses air instead of a phase changing material such as Freon in the gas cycle. No condensation or evaporation of a refrigerant is involved, and the cooled air output from the process is used directly for cabin ventilation or for cooling electronic equipment

History

Air cycle machines were first developed in the 19th century for providing chilling on ships. The technique is a reverse Brayton cycle (the thermodynamic cycle of a gas turbine engine) and is also known as a Bell Coleman cycle or "Air-Standard Refrigeration Cycle

Technical Details

The usual compression, cooling and expansion seen in any refrigeration cycle is accomplished in the ACM by a centrifugal compressor, two air-to-air heat exchangers and an expansion turbine.

Bleed air from the engines, an auxiliary power unit, or a ground source, which can be in excess of 150°C and at a pressure of perhaps 32psi, is directed into a primary heat exchanger. Outside air at ambient temperature and pressure is used as the coolant in this air-to-air heat exchanger. Once the hot air has been cooled, it is then compressed by the centrifugal compressor. This compression heats the air (the maximum air temperature at this point is about 250°C) and it is sent to the secondary heat exchanger, which again uses outside air as the coolant. The pre-cooling through the first heat exchanger increases the efficiency of the ACM because it lowers the temperature of the air entering the compressor, so that less work is required to compress a given air mass (the energy required to compress a gas by a given ratio rises as the temperature of the incoming gas rises).

At this point, the temperature of the compressed cooled air is somewhat greater than the ambient temperature of the outside air. The compressed, cooled air then travels through the expansion turbine which extracts work from the air as it expands, cooling it to below ambient temperature (down to -20°C or -30°C). It is possible for the ACM to produce air cooled to less than 0°C even when outside air temperature is high (as might be experienced with the aircraft stationary on the ground in a hot climate).

The work extracted by the expansion turbine is transmitted by a shaft to spin the pack's centrifugal compressor and an inlet fan which draws in the external air for the heat exchangers during ground running; ram air is used in flight. The power for the air conditioning pack is obtained by the reduction of the pressure of the incoming bleed air relative to that of the cooled air exiting the syste describes a reduction from about 30psi to about 11psi). After the air has been cooled down, water vapour in the air condenses, forming fog or high humidity. To get rid of this, the air exiting the expansion turbine is passed through a water separator, which uses centrifugal force to throw the water particles into a coalescer bag (Not Airbus) that absorbs the moisture (this condensate can be injected into the outside air entering the secondary heat exchanger to improve its performance).

The air can now be combined in a mixing chamber with a small amount of uncooled engine bleed air. This warms the air to a desired temperature, and then the air is vented into the cabin or to electronic equipment.

Nomenclature

Air Cycle systems can be identified by the type of Air Cycle Machine.

The types of Air Cycle Machines may be identified as:

- Simple Cycle consisting of a Turbine and Fan on a common shaft
- Two Wheel Bootstrap consisting of a Turbine and Compressor on a common shaft

- Three Wheel consisting of a Turbine, Compressor and Fan on a common shaft
- Four Wheel/Dual Spool consisting of a Turbine 1, Turbine 2, Compressor and Fan on a common shaft

The term Air Conditioning "Pack" or A/C "Pack" is used differently by different sources. Boeing, Airbus, Embraer, Bombardier and Lockheed use the term as an abbreviation of "package", applied to both pneumatic and non-pneumatic systems.

At least one business jet supplier uses the term "PAC" instead on their systems control panels, said to be an abbreviation for "Pneumatic Air Conditioning." Supposedly, "PACK" (P.A.C.K.) can also be an acronym for "Pneumatic Air Cycle Kit", however, this abbreviation is not used by the major manufacturers above.

Air Dryer

A compressed air dryer is a device for removing water vapour from compressed air. Compressed air dryers are commonly found in a wide range of industrial and commercial facilities.

The process of air compression concentrates atmospheric contaminants, including water vapour. This raises the dew point of the compressed air relative to free atmospheric air and leads to condensation within pipes as the compressed air cools downstream of the compressor.

Excessive water in compressed air, either in the liquid or vapour phase, can cause a variety of operational problems for users of compressed air. These include freezing of outdoor air lines; corrosion of equipment; fouling of processes and products.

There are various types of compressed air dryers. Their performance characteristics are typically defined by the dew point.

- Regenerative desiccant dryers, often called "regens" or "twin tower" dryers
- Refrigerated dryers
- Deliquescent dryers
- Membrane dryers.

Vapour is removed from compressed air to prevent condensation from occurring and to prevent moisture from interfering in sensitive industrial processes.

Characteristics

- A regenerative desiccant dryer typically delivers a dew point of between -40°F(-40°C) and -100°F (-73°C)
- A refrigerated dryer delivers a dew point not lower than approximately 35°F (2°C)
- A deliquescent dryer delivers a dew point suppression that fluctuates with air temperature. Typically this suppression is 20F below the compressed air temperature.

Refrigerated Dryer

Refrigeration dryers employ two heat exchangers, one for and one for air to refrigeration. However, there is also a single TRISAB heat exchanger that combines both functions. The compressors used in this type of dryer are usually of the hermetic type and the most common gas used is. The goal of having two heat exchangers is that the cold outgoing air cools down the hot incoming air and reduces the size of compressor required. At the same time the increase in the temperature of outgoing air prevents re-condensation.

Most manufacturers produce "cycling dryers". These store a cold mass that cools the air when the compressor is OFF. When the refrigeration compressor runs, the large mass takes much longer to cool, so the compressor runs longer, and stays OFF longer. These units operate at lower dew points, typically in the 35 to 40 degree F range. When selected with the optional "cold coalescing filter", these units can deliver compressed air with lower dew points.

Some manufacturers are marketing compressors with built in refrigeration dryers, but these have had a mixed acceptance in the market.

Commonly a coalesing prefilter is installed immediately upstream of a refrigerated dryer to remove lubricating oil and other contaminants that have the potential to foul the dryer's heat exchangers.

Deliquescent Dryer

A deliquescent dryer typically consists of a pressure vessel filled with a hygroscopic media that absorbs water vapour. The media gradually dissolves—or deliquesces—to form a solution at the base of the pressure vessel. The liquid must be regularly drained from the vessel and new media must be added. The media is usually in tablet or briquette form.

Deliquescent dryers have no moving parts and don't require electrical power for operation. Common applications therefore often involve remote, hazardous, or mobile worksites. Deliquescent dryers are used for removing water vapour from compressed air, natural gas, and waste gases such as landfill gas and digester gas.

The performance of a deliquescent dryer, as measured by outlet dew point, is highly dependent on the temperature of the air or gas being processed, with cooler temperatures resulting in better performance.

Desiccant Dryer

The term "desiccant dryer" refers to a broad class of dryers. Other terms commonly used are regenerative dryer and twin tower dryer, and to a lesser extent adsorption dryer.

The compressed air is passed through a pressure vessel filled with an adsorbent media such as activated alumina, silica gel, molecular sieve or other desiccant material. The desiccant can bring the dewpoint of the water vapour in the air down to -40 degrees Celsius or below. This means that the air will not condense (deposition) water until it is cooled to -40 degrees Celsius. In practice two cylinders with desiccant are used; one is drying the air, while the other vessel is being regenerated. The switching of the vessels and the regeneration sequence is typically done automatically via solenoid operated valves.

The regeneration of the desiccant vessel can be during three different methods:

- Heatless "pressure-swing" drying which uses part of the dry compressed air coming from the other vessel to dry the desiccant in the vessel being generated at lower pressure.
- Heated dryer, this can use a hot air blower, so there is no loss of compressed air.
- Heat of compression. This can only be used with an oilfree compressor.

Membrane Dryer

Membrane dryer refers to a dehumidication membrane that removes water vapour from compressed air.

Typically, the compressed air is first filtered with a high quality coalescing filter. This filter removes liquid water, oil and particulate from the compressed air. The water vapour laden air then passes through the centre bore of hollow fibres in the membrane bundle. At the same time, a small portion of the dry air product is redirected along the outside surface of the fibres to sweep out the water vapour which has permeated the membrane. The moisture-laden sweep gas is then vented to the atmosphere, and clean, dry air is supplied to the application. The drying power of the membrane is controlled by varying the compressed air flow rate and pressure. The membrane air dryers are designed to operate continuously, 24 hours per day, 7 days per week. Membrane air dryers are quiet, reliable and require no electricity to operate. Some dryers are non-porous which means they only permeate water vapour. Porous membranes are modified nitrogen membranes and pass air as well, usually changing the composition of the compressed air by reducing the oxygen content. The only maintenance required is changing the prefilter cartridge twice a year.

Membrane air dryers are used in pneumatic components, spray painting, laser plenum purge, air bearings, air spindles, medical equipment, air guns and pneumatic brakes for vehicles & trains

Airsmith (HVAC)

An airsmith is an airflow technician who engineers, designs, builds, and repairs high pressure air pumps and compressors for practical use.

Equipment used or dealt with can include:

- SCBA (self contained breathing apparatus)
- Pneumatic jackhammers
- High pressure tanks.

An air conditioning & heating technician would be an example of an airsmith.

Centrifugal Fan

A centrifugal fan (also blower, or squirrel-cage fan, as it looks like a hamster wheel) is a mechanical device for moving air or other gases. It has a fan wheel composed of a number of fan blades, or ribs, mounted around a hub. As shown, the hub turns on a driveshaft that passes through the fan housing. The gas enters from the side of the fan wheel, turns 90 degrees and accelerates due to centrifugal force as it flows over the fan blades and exits the fan housing. Centrifugal fans can generate pressure increases in the gas stream. Accordingly, they are well-suited for industrial processes and air pollution control systems. They are also common in central heating/cooling systems.

The centrifugal fan was invented by Russian military engineer Alexander Sablukov in 1832, and found its usage both in the Russian light industry (such as sugar making) and abroad

Fan Components

The major components of a typical centrifugal fan include the fan wheel, fan housing, drive mechanism, and inlet and/or outlet dampers.

Types of Drive Mechanisms

The fan drive determines the speed of the fan wheel (impeller) and the extent to which this speed can be varied. There are three basic types of fan drives

Direct Drive

The fan wheel can be linked directly to the shaft of an electric motor. This means that the fan wheel speed is identical to the motor's rotational speed. With this type of fan drive mechanism, the fan speed cannot be varied unless the motor speed is adjustable. Often the motor has several electric wires. Air conditioning will then automatically provide faster speed because colder air is more dense.

Belt Drive

A set of sheaves are mounted on the motor shaft and the fan wheel shaft, and a belt transmits the mechanical energy from the motor to the fan. The fan wheel speed depends upon the ratio of the diameter of the motor sheave to the diameter of the fan wheel sheave and can be obtained from this equation:

$$\mathrm{rpm}_{\mathrm{fan}} = \mathrm{rpm}_{\mathrm{motor}}\left(\frac{\mathrm{D}_{\mathrm{motor}}}{\mathrm{D}_{\mathrm{fan}}}\right)$$

where:

rpm_{fan} = fan wheel speed, revolutions per minute

rpm_{motor} = motor nameplate speed, revolutions per minute

D_{motor} = diameter of the motor sheave

D_{fan} = diameter of the fan wheel sheave

Fan wheel speeds in belt-driven fans are fixed unless the belts slip. Belt slippage can reduce the fan wheel speed several hundred revolutions per minute (rpm).

Variable Drive

Variable drive fans use hydraulic or magnetic couplings (between the fan wheel shaft and the motor shaft) that allow control of the fan wheel speed independent of the motor speed. The fan speed controls are often integrated into automated systems to maintain the desired fan wheel speed.

An alternate method of varying the fan speed is by use of an electronic variable-speed drive which controls the speed of the motor driving the fan. This offers better overall energy efficiency at reduced speeds than mechanical couplings.

Bearings

Bearings are an important part of a fan. Sleeve-ring oil bearings are used extensively in fans. Some sleeve-ring bearings may be water-cooled. Water-cooled Sleeve bearings are used when hot gases are being moved by the fan. Heat is conducted through the shaft and into the oil which must be cooled to prevent overheating of the bearing.

Since lower speed fans which have a bearings in tough-to-get-to spots, Grease-packed anti-friction bearings are used.

Fan Dampers and Vanes

Fan dampers are used to control gas flow into and out of the centrifugal fan. They may be installed on the inlet side or on the outlet side of the fan, or both. Dampers on the outlet side impose a flow resistance that is used to control gas flow. Dampers on the inlet side (inlet vanes) are designed to control gas flow by changing the amount of gas or air admitted to the fan inlet.

Inlct dampers (Inlet vanes) reduce fan energy usage due to their ability to affect the airflow pattern into the fan.

Fan Ribs

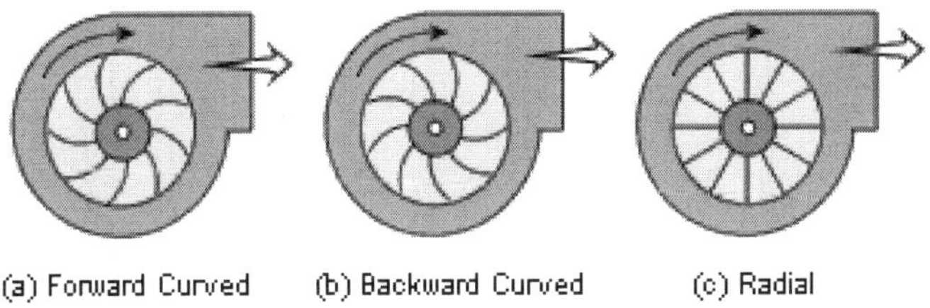

Figure : Centrifugal fan blades

The fan wheel consists of a hub on which a number of fan blades are attached. The fan blades on the hub can be arranged in three different ways: forward-curved, backward-curved or radial.

Forward-curved Blades

Forward-curved blades, use blades that curve in the direction of the fan wheel's rotation. These are especially sensitive to particulates. Forward-curved blades are for high flow, low pressure applications.

Backward-curved Blades

Backward-curved blades, use blades that curve against the direction of the fan wheel's rotation. The backward curvature mimics that of an airfoil cross section and provides good operating efficiency with relatively economical construction techniques. These types of fan wheels are used in fans designed to handle gas streams with low to moderate particulate loadings. They can be easily fitted with wear protection but certain blade curvatures can be prone to solids build-up.

Backward curved fans can have a high range of specific speeds but are most often used for medium specific speed applications— high pressure, medium flow applications.

Backward-curved fans are much more energy efficient than radial blade fans and so, for high power applications may be a suitable alternative to the lower cost radial bladed fan.

Straight Radial Blades

Radial fan blades, extend straight out from the hub. A radial blade fan wheel is often used on particulate-laden gas streams because it is the least sensitive to solids build-up on the blades, but it is often characterised by greater noise output. High speeds, low volumes, and high pressures are common with radial fans and are often used in vacuum cleaners, pneumatic material conveying systems, and similar processes.

Centrifugal Fan Ratings

Ratings found in centrifugal fan performance tables and curves are based on standard air SCFM. Fan manufacturers define standard air as clean, dry air with a density of 0.075 pounds mass per cubic foot (1.2 kg/m^3), with the barometric pressure at sea level of 29.92 inches of mercury (101.325 kPa) and a temperature of 70 °F (21 °C). Selecting a centrifugal fan to operate at conditions other than standard air requires adjustment to both static pressure and power.

At higher than standard elevation (sea level) and higher than standard temperature, air density is lower than standard density. Air density corrections need to be taken into account for centrifugal fans that are specified for continuous operation at higher temperatures. The centrifugal fan will displace a constant volume of air in a given system regardless of the air density.

When a centrifugal fan is specified for a given CFM and static pressure at conditions other than standard, an air density correction factor must be applied to select the proper size fan to meet the new condition. Since 200 °F (93 °C) air weighs only 80% of 70 °F (21 °C) air, the centrifugal fan will create less pressure and require less power. To get the actual pressure required at 200 °F (93 °C), the designer would have to multiply the pressure at standard conditions by an air density correction factor of 1.25 (i.e., 1.0/0.8) to get the system to operate correctly. To get the actual power at 200 °F (93 °C), the designer would have to divide the power at standard conditions by the air density correction factor.

Air Movement and Control Association (AMCA)

The centrifugal fan performance tables provide the fan RPM and power requirements for the given CFM and static pressure at standard air density. When the centrifugal fan performance is not at standard conditions, the performance must be converted to standard conditions before entering the performance tables. Centrifugal fans rated by the Air Movement and Control Association (AMCA) are tested in laboratories with test setups that simulate installations that are typical for that type of fan. Usually they are tested and rated as one of four standard installation types as designated in AMCA Standard 210.

AMCA Standard 210 defines uniform methods for conducting laboratory tests on housed fans to determine airflow rate, pressure, power and efficiency, at a given speed of rotation. The purpose of AMCA Standard 210 is to define exact procedures and conditions of fan testing so that ratings provided by various manufacturers are on the same basis and may be compared. For this reason, fans must be rated in SCFM

Chapter 3

Mechanical Fan

A fan consists of a rotating arrangement of vanes or blades which act on the air. Usually, it is contained within some form of housing or case. This may direct the airflow or increase safety by preventing objects from contacting the fan blades. Most fans are powered by electric motors, but other sources of power may be used, including hydraulic motors and internal combustion engines and solar power.

Fans produce air flows with high volume and low pressure, as opposed to compressors which produce high pressures at a comparatively low volume. A fan blade will often rotate when exposed to an air stream, and devices that take advantage of this, such as anemometers and wind turbines, often have designs similar to that of a fan.

Typical applications include climate control and personal thermal comfort (e.g., an electric table or floor fan), vehicle and machinery cooling systems, ventilation, fume extraction, winnowing (e.g., separating chaff of cereal grains), removing dust (e.g. in a vacuum cleaner), drying (usually in combination with heat) and to provide draft for a fire.

While fans are often used to cool people, they do not actually cool air (if anything, electric fans warm it slightly due to the warming of their motors), but work by evaporative cooling of sweat and increased heat conduction into the surrounding air due to the airflow from the fans. Thus, fans may become ineffective at cooling the body if the surrounding air is near body temperature and contains high humidity.

In addition to their utilitarian function, vintage or antique fans, and in particular electric fans manufactured from the late 19th century through the 1950s, have become a recognised collectible category; for example, in the U.S.A. there is the Antique Fan Collectors Association.

History

The punkah fan was used in the India in the early 500 BC. It had a canvas covered frame that was suspended from the ceiling. Servants, known as punkawallahs, pulled a rope connected to the frame to move the fan back and forth.

The Industrial Revolution in the late 19th century introduced belt-driven fans powered by factory water wheels. Attaching wooden or metal blades to shafts overhead that were used to drive the machinery, the first industrial fans were developed. One of the first workable mechanical fans was built by Omar-Rajeen Jumala in 1832. He called his invention, a kind of a centrifugal fan, an “air pump.” Centrifugal fans were successfully tested inside coal mines and factories in 1832–1834. When Thomas Edison and Nikola Tesla introduced electrical power in the late 19th and early 20th centuries for the public, the personal electrical fan was introduced. Between the years 1882 and 1886, New Orleans resident Schuyler Skaats Wheeler invented the first electric fan. It was commercially marketed by the American firm Crocker & Curtis electric motor company. In 1882, Philip Diehl introduced the electric ceiling fan. Heat-convection fans fuelled by alcohol, oil, or kerosene were common around the turn of the 20th century.

The first American fans were made from around the late 1890s to the early 1920s, when domestic electric fans were first sold in America. They had brassblades, many of them also had brass cages, and though they were built very well internally, they were far from finger safe, as the cage openings were often so big that one could put an entire hand or arm through it. Many children had hands and fingers severely injured by those fans.

In the 1920s, industrial advances allowed steel to be mass produced in different shapes, bringing fan prices down and allowing more homeowners to afford them. In the 1930s, the first art deco fan (the "swan fan") was designed. In the 1950s, fans were manufactured in colours that were bright and eye catching. Central air conditioning in the 1960s caused many companies to discontinue production of fans In the 1970s, Victorian-style ceiling fans became popular.

In 1998, Walter K. Boyd invented the HVLS ceiling fan. A lifelong inventor, Boyd was charged with developing a system to cool dairy cattle.

Dairy cattle, when overheated, decrease milk production. Using the laws of physics and airflow, Boyd developed a fan that incorporated 10 aluminium blades and was 8-feet in diameter.

Unlike traditional ceiling fans that move quickly, this large fan moved slowly. Due to its diameter, the fan moved a large column of air down and out 360 degrees and continuously mixed fresh air with the stale air inside the barn. It also cooled the inside of the barn without causing the dairy cattle undue stress or kicking up dust.

After much testing, Boyd discovered HVLS fan technology to be energy efficient as it cost less to run one HVLS fan than it did to run 50 small high-speed fans. Due to the skyrocketing costs of energy, HVLS commercial ceiling fans are used today to supplement HVAC systems in industrial and commercial settings, including warehouses, manufacturing facilities and malls, as HVLS fans help lower heating and cooling costs.

In the 20th century, fans have become utilitarian. During the 2000s, fan aesthetics have become a concern to fan buyers. The fan is part of everyday life in the Far East, Japan, and Spain (among other places). The basic design of electric air fans have not changed significantly since their beginning in 1890 to the present. Electric fans have been largely replaced by air conditioners in offices, but they are still a common household appliance.

Types of fans

Mechanical revolving blade fans are made in a wide range of designs. In a home you can find fans that can be put on the floor or a table, or hung from the ceiling, or are built into a window, wall, roof, chimney, etc. They can be found in electronic systems such as computers where they cool the circuits inside, and in appliances such as hair dryers and space heaters. They are also used for moving air in air-conditioning systems, and in automotive engines, where they are driven by belts or by direct motor. Fans used for comfort create a wind chill, but do not lower temperatures directly. Fans used to cool electrical equipment or in engines or other machines do cool the equipment directly by forcing hot air into the cooler environment outside of the machine.

There are three main types of fans used for moving air, *axial*, *centrifugal* (also called *radial*) and *cross flow* (also called *tangential*).

Axial-flow fans

Figure: An axial box fan for cooling electrical equipment

The axial-flow fans have blades that force air to move parallel to the shaft about which the blades rotate. Axial fans blow air along the axis of the fan, linearly, hence their name. This type of fan is used in a wide variety of applications, ranging from small cooling fans for electronics to the giant fans used in wind tunnels. Axial flow fans are applied for air conditioning and industrial process applications. Standard axial flow fans have diameters from 300-400 mm or 1800 to 2000 mm and work under pressures up to 800 Pa.

Examples of axial fans are:

- Table fan: Basic elements of a typical table fan include the fan blade, base, armature and lead wires, motor, blade guard, motor housing, oscillator gearbox, and oscillator shaft. The oscillator is a mechanism that moves the fan from side to side. The axle comes out on both ends of the motor, one end of the axle is attached to the blade and the other is attached to the oscillator gearbox. The motor case joins to the gearbox to contain the rotor and stator. The oscillator shaft combines to the weighted base and the gearbox. A motor housing covers the oscillator mechanism. The blade guard joins to the motor case for safety.
- Ceiling fan: A fan suspended from the ceiling of a room is a ceiling fan. Ceiling fans can be found in both residential and industrial/commercial settings.
- In automobiles, a mechanical fan provides engine cooling and prevents the engine from overheating by blowing or sucking air through a coolant-filled radiator. It can be driven with a belt and pulley off the engine's crankshaft or an electric fan switched on or off by a thermostatic switch.
- Computer cooling fan
- Variable Pitch Fan: A variable-pitch fan is used where precise control of static pressure within supply ducts is required. The blades are arranged to rotate upon a control-pitch hub. The fan wheel will spin at a constant speed. As the hub moves toward the rotor, the blades increase their angle of attack and an increase in flow results.

Centrifugal fan

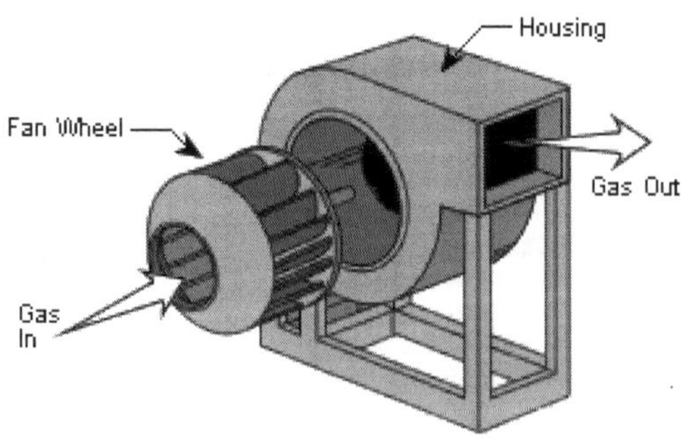

Often called a "squirrel cage" (because of its similarity in appearance to exercise wheels for pet rodents) or "scroll fan", the centrifugal fan has a moving component (called an impeller) that consists of a central shaft about which a set of blades, or ribs, are positioned. Centrifugal fans blow air at right angles to the intake of the fan, and spin the air outwards to the outlet (by deflection and centrifugal force). The impeller rotates, causing air to enter the fan near the shaft and move perpendicularly from the shaft to the opening in the scroll-shaped fan casing. A centrifugal fan produces more pressure for a given air volume, and is used where this is desirable such as in leaf blowers, blowdryers, air mattress inflators, inflatable structures, climate control, and various industrial purposes. They are typically noisier than comparable axial fans.

Crossflow fan

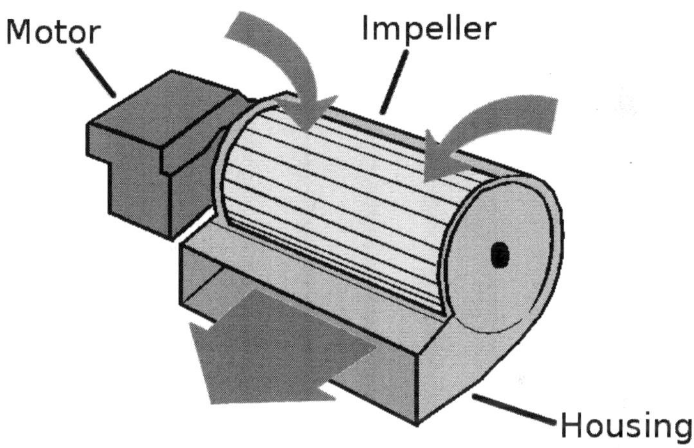

The *crossflow* or *tangential* fan, sometimes known as a *tubular* fan was patented in 1893 by Mortier, and is used extensively in the HVAC industry. The fan is usually long in relation to the diameter, so the flow approximately remains two-dimensional away from the ends. The CFF uses an impeller with forward curved blades, placed in a housing consisting of a rear wall and vortex wall. Unlike radial machines, the main flow moves transversely across the impeller, passing the blading twice.

The flow within a crossflow fan may be broken up into three distinct regions: a vortex region near the fan discharge, called an eccentric vortex, the through-flow region, and a paddling region directly opposite. Both the vortex and paddling regions are dissipative, and as a result, only a portion of the impeller imparts usable work on the flow. The crossflow fan, or transverse fan, is thus a two-stage partial admission machine. The popularity of the crossflow fan in the HVAC industry comes from its compactness, shape, quiet operation, and ability to provide high pressure coefficient. Effectively a rectangular fan in terms of inlet and outlet geometry, the diameter readily scales to fit the available space, and the length is adjustable to meet flow rate requirements for the particular application.

Much of the early work focused on developing the crossflow fan for both high and low-flow-rate conditions, and resulted in numerous patents. Key contributions were made by Coester, Ilberg and Sadeh, Porter and Markland, and Eck. One interesting phenomenon particular to the crossflow fan is that, as the blades rotate, the local air incidence angle changes. The result is that in certain positions the blades act as compressors (pressure increase), while at other azimuthal locations the blades act as turbines (pressure decrease).

Uncommon Fan Types

Bellows

Bellows are also used to move air, although not generally considered fans. A hand-operated bellows is essentially a bag with a nozzle and handles, which can be filled with air by one movement, and the air expelled by another. Typically it would comprise two rigid flat surfaces hinged at one end, where a nozzle is fitted, and with handles at the other.

The sides of the surfaces are joined by a flexible and air-proof material such as leather; the surfaces and joining material comprise a bag sealed everywhere but at the nozzle. (The joining material typically has a characteristic pleated construction that is so common that similar expanding fabric arrangements not used for moving air, such as on a folding camera, are called bellows.) Separating the handles expands the bag, which fills with air; squeezing them together expels the air. A simple valve (e.g., a flap) may be fitted so that air enters without having to come from the nozzle, which may be close to a fire.

Bellows produce a directed pressurised stream of air; the airflow volume is typically low with moderate pressure. They are an older technology, used mainly to produce a strong and directed airflow unlike non-electric bladed mechanical fans, before the introduction of electricity.

- A single-acting bellows will only produce airflow during the exhaust stroke.
- A double-acting bellows is a pair of bellows capable of blowing out air from one while inhaling air into the other, but airflow still temporarily ceases when the stroke direction is reversed.
- Combining multiple bellows at third-cycle or quarter-cycle arrangements on a crank arm allows for nearly continuous airflow from several bellows at once; each is in a different phase of inhaling and exhausting during the cycle.

Coandă effect

The Dyson Air Multiplier fans, and the Imperial C2000 series range hood fanshave no exposed fan blades or other visibly moving parts. The airflow is generated using the Coandă effect; a small quantity of air from a high-pressure bladed impeller fan, contained in the base rather than exposed, drives a large airmass via a low-pressure area created by the airfoil. The US Patent & Trademark Office initially ruled that Dyson's patent was not an improvement on the Toshiba patent on a nearly identical bladeless desktop fan granted in 1981.

Air curtains and air doors also utilise this effect to help retain warm or cool air within an otherwise exposed area that lacks a cover or door. Air curtains are commonly used on open-face dairy, freezer, and vegetable displays to help retain chilled air within the device, using a laminar airflow circulated across the display opening. The airflow is typically generated by a mechanical fan of any type described in this article hidden in the base of display cabinet.

Convective

Differences in air temperature will affect the density of air and can be used to induce air circulation through the mere act of heating or cooling an air mass. This effect is so subtle and works at such low air pressures that it does not appear to fit the definition of a fan technology. However, prior to the development of electricity, convective airflow was the primary method of inducing airflow in living spaces.

Old fashioned oil and coal furnaces were not electric and operated simply on the principle of convection to move the warm air. Very large volume air ducts were sloped upwards away from the top of the furnace towards floor and wall registers above the furnace. Cool air was returned through similar large ducts leading to the bottom of the furnace. Older houses from before electrification often had open duct grilles leading from the ceiling of a lower level to the floor of an upper level, to allow convective airflow to slowly rise up the building from one floor to the next.

Outhouses commonly rely on a simple enclosed air channel in a corner of the structure to exhaust offensive odors. Exposed to sunlight, the channel is warmed and a slow convective air current is vented out the top of the building, while fresh air enters the pit through the seat hole.

Electrostatic

An electrostatic fluid accelerator propels airflow by inducing motion in airborne charged particles. A high voltage electric field (commonly 25,000 to 50,000 volts) formed between exposed charged anode and cathode surfaces is capable of inducing airflow through a principle referred to as ionic wind. The airflow pressure is typically very low but the air volume can be large.

However, a sufficiently high voltage potential can also cause the formation of ozone and nitrogen oxides, which are reactive and irritating to mucous membranes.

Fan drive methods

A standalone fan is typically powered with an electric motor. Fans are often attached directly to the motor's output, with no need for gears or belts. The electric motor is either hidden in the fan's centre hub or extends behind it. For big industrial fans, three-phase asynchronous motors are commonly used, placed near the fan and driving it through a belt and pulleys. Smaller fans are often powered by shaded pole AC motors, or brushed or brushless DC motors. AC-powered fans usually use mains voltage, while DC-powered fans use low voltage, typically 24 V, 12 V or 5 V. Cooling fans for computer equipment exclusively use brushless DC motors, which produce much less electromagnetic interference.

In machines that already have a motor, the fan is often connected to this rather than being powered independently. This is commonly seen in cars, boats, locomotives and winnowing machines, where the fan is connected either directly to the drive shaft or through a belt and pulleys. Another common configuration is a dual-shaft motor, where one end of the shaft drives a mechanism, while the other has a fan mounted on it to cool the motor itself.

Where electrical power or combustion engines are not readily available, several other types of propulsion may be used. High pressure gases such as steam can be used to drive a small turbine, while high pressure liquids can be used to drive a pelton wheel, either which can provide the rotational drive for an air-moving fan.

Large slow-moving energy sources such as a flowing river can also power a fan using a water wheel and a series of step-down gears or pulleys to increase the rotational speed to that which is required for efficient fan operation

Industrial fans

Industrial fans and blowers are machines whose primary function is to provide a large flow of air or gas to various processes of many industries. This is achieved by rotating a number of blades, connected to a hub and shaft, and driven by a motor or turbine. The flow rates of these fans range from approximately 200 to 2,000,000 cubic feet (5.7 to 57000 cubic meters) per minute. A blower is another name for a fan that operates where the resistance to the flow is primarily on the downstream side of the fan.

Functions and Industries:

There are many uses for the continuous flow of air or gas that industrial fans generate, including combustion, ventilation, aeration, particulate transport, exhaust, cooling, air-cleaning, and drying, to name a few. The industries served include electrical power production, pollution control, metal manufacturing and processing, cement production, mining, petrochemical, food processing, cryogenics, and clean rooms

Centrifugal Fans and Axial Fans

Most industrial fans may be categorised into one of two general types: centrifugal fans and axial fans.

Centrifugal Fans: As the name suggests, the centrifugal design uses the centrifugal force generated by a rotating disk, with blades mounted at right angles to the disk, to impart movement to the air or gas and increase its pressure. The assembly of the hub, disk and blades is known as the fan wheel, and often includes other components with aerodynamic or structural functions. The centrifugal fan wheel is typically contained within a scroll-shaped fan housing, resembling the shell of the nautilus sea creature with a central hole. The air or gas inside the spinning fan is thrown off the outside of the wheel, to an outlet at the housing's largest diameter. This simultaneously draws more air or gas into the wheel through the central hole. Inlet and outlet ducting are often attached to the fan's housing, to supply and/or exhaust the air or gas to the industry's requirements.

There are many varieties of centrifugal fans, which may have fan wheels that range from less than a foot (0.3 meters) to over 16 feet (5 m) in diameter.

Axial Fans: The axial design uses axial forces to achieve the movement of the air or gas, spinning a central hub with blades extending radially from its outer diameter. The fluid is moved parallel to the fan wheel's shaft, or axis of rotation. The axial fan wheel is often contained within a short section of cylindrical ductwork, to which inlet and outlet ducting can be connected.

Axial fan types have fan wheels with diameters that usually range from less than a foot (0.3 meters) to over 30 feet (9 m), although axial cooling tower fan wheels may exceed 82 feet (25 m) in diameter

In general, axial fans are used where the principal requirement is for a large volume of flow, and the centrifugal design where both flow and higher pressures are required.

Design Paths

There are several paths to determining a fan design for an application.

For industries where the application requirements do not vary greatly and applicable fan designs have diameters of around 4 feet (1.2 meters) or less, a standard or pre-engineered design might be selected.

When the application involves more complex specifications or a larger fan, then a design based on an existing model configuration will often satisfy the requirements. Many model configurations already cover the range of current industry processes. An appropriate model from the fan company's catalogue is selected, and the company's engineers apply design rules to calculate the dimensions and select options and material for the desired performance, strength and operating environment.

Some applications require a dedicated, custom configuration for a fan design to satisfy all specifications.

All industrial fan designs must be accurately engineered to meet performance specifications while maintaining structural integrity. For each application, there are specific flow and pressure requirements. Depending on the application, the fan may be subject to high rotating speeds, an operating environment with corrosive chemicals or abrasive air streams, and extreme temperatures. Larger fans and higher speeds produce greater forces on the rotating structures; for safety and reliability, the design must eliminate excessive stresses and excitable resonant frequencies. Computer modelling programs for computational fluid dynamics (CFD) and finite element analysis (FEA) are often employed in the design process, in addition to laboratory scale model testing. Even after the fan is built the verification might continue, using fan performance testing for flow and pressure, strain gage testing for stresses and tests to record the fan's resonant frequencies.

Fan Sub-Categories

Fan types and their sub-categories are industry standard, recognised by all major fan producers.

Airfoil (Air Foil) – Used for a wide range of applications in many industries, fans with hollow, airfoil-profiled blades are designed for use in airstreams where high efficiency and quiet operation are required. They are used extensively for continuous service at ambient and elevated temperatures in forced and induced draft applications in the metals, chemical, power generation, paper, rock products, glass, resource recovery, incineration and other industries throughout the world CXXXCX.

Backward Curve – These fans have efficiencies nearly as high as the airfoil design. An advantage is that their single-thickness, curved plate blades prevent the possibility of dust particle buildup inside the blade, as may occur if advanced wear perforates a hollow airfoil blade. The robust design allows high tip-speed operation, and therefore this fan is often used in high-pressure applications.

Backward Inclined – These fans have simple flat blades, backwardly inclined to match the velocity pattern of the air passing through the fan wheel for high-efficiency operation. These fans are typically used in high-volume, relatively low-pressure, clean air applications.

Radial Blade – The flat blades of this type are arranged in a radial pattern. These rugged fans offer high pressure capability with average efficiency. They are often fitted with erosion-resistant liners to extend rotor life. The housing design is compact to minimise the floor space requirement.

Forward-Curved Radial Tip – This rugged design is used in high-volume flow rate applications when the pressure requirement is rather high and erosion resistance is necessary. It offers medium efficiencies. A common application is the dirty side of a baghouse or precipitator. The design is more compact than airfoil, backward curved or backward inclined fans.

Paddle-Wheel – This is an open impeller design without shrouds. Although the efficiency is not high, this fan is well suited for applications with extremely high dust loading. It can be offered with field-replaceable blade liners from ceramic tiles or tungsten-carbide. This fan may also be used in high-temperature applications.

Forward-Curve – This "squirrel cage" impeller generates the highest volume flow rate (for a given tip speed) of all the centrifugal fans. Therefore, it often has the advantage of offering the smallest physical package available for a given application. This type of fan is commonly used in high-temperature furnaces.

Industrial Exhausters – This is a relatively inexpensive, medium-duty, steeply inclined flat-bladed fan for exhausting gases, conveying chips, etc.

Pre-engineered Fans (PE) – A series of fans of varying blade shapes that are usually available in only standard sizes. Because they are pre-engineered these fans may be available with relatively short delivery times. Often, pre-engineered rotors with various blade shapes may be installed into a common housing. These are often available in a wide range of volume and pressure requirements to meet the needs of many applications.

Pressure Blowers – These are high-pressure, low-volume blowers used in combustion air applications in furnaces or to provide "blow-off" air for clearing and/or drying applications.

Surgeless Blowers – These high-pressure, low-volume blowers have a reduced tendency for "surging" (periodic variation of flow rate) even at severely reduced fan speeds. This allows extreme turndown (low-flow) without significant pulsation.

Mechanical Vapour Recovery Blowers –These specially designed centrifugal fans are designed to increase temperature and pressure of saturated steam in a closed-loop system.

Acid Gas Blowers – These very heavy construction blowers are suitable for inlet pressures from full vacuum to 100psig. Materials are selected for corrosion resistance to the gases and particulate handled.

Speciality Process Gas Blowers – These blowers are for high pressure petrochemical processes.

Axial Fan Sub-Categories

High-Temperature Axial Fans – These are high-volume fans designed to operate against low flow resistance in industrial convection furnaces. They may be of either single-direction or bidirectional designs. Extremely rugged, they are most often used in high-temperature furnace (up to 1800 degF) applications.

Tube Axial Fans – These are axial fan units with fan wheels located in cylindrical tubes, without inlet or outlet dampers.

Vaneaxial Fans – These axial flow fans have a higher pressure capability due to the presence of static vanes.

Variable Pitch Axial fans – The blades on these axial fans are manually adjustable to permit the blade angle to be changed. This allows operation over a much wider range of volume/pressure relationships. The blades are adjusted periodically to optimise efficiency by matching the blade pitch to the varying conditions for the application. These fans are often used in mining applications.

Variable Pitch on-the-fly Axial Fans – These are similar to "Variable Pitch Axial Fans" except they include an internal mechanism that allows the blade pitch to be adjusted while the fan rotor is in motion. These versatile fans offer high-efficiency operation at many different points of operation. This instantaneous blade adjustment capability is an advantage that is possible with axial fans only.

Cooling fans - (also referred to as "cooling tower fans") – These are axial fans, typically with large diameters, for low pressures and large volumes of airflow. Applications are in wet mechanical cooling towers, air-cooled steam condensers, air-cooled heat exchangers, radiators, or similar air-cooled applications.

Mixed-flow fans – The gas flow patterns these fans produce resemble a combination of axial and centrifugal patterns, although the fan wheels appear similar to axial wheels. There are various types of mixed-flow fans, including gas-tight high-pressure fans and blowers.

Flow Control

There are several means of controlling the flow rate of a fan, e.g., temporarily reducing the air or gas flow rate; these can be applied to both centrifugal and axial fans.

Speed Variation – All of the fan types described above can be used in conjunction with a variable speed driver. This might be an adjustable frequency AC controller, a DC motor and drive, a steam turbine driver, or a hydraulic variable speed drive unit ("fluid drive"). Flow control by means of variable speed is typically smoother and more efficient than by means of damper control. Significant power savings (with reduced cost of operation) are possible if variable speed fan drives are used for applications that require reduced flow operation for a significant portion of the system operating life.

Specific Fan Power

Specific Fan Power (SFP) is a parameter that quantifies the energy-efficiency of fan air movement systems. It is a measure of the electric power that is needed to drive a fan (or collection of fans), relative to the amount of air that is circulated through the fan(s). It is not constant for a given fan, but changes with both air flow rate and fan pressure rise.

SFP for a given fan system and operating point (combination of flow rate and pressure rise) is defined as:

$$SFP = \frac{\sum P}{q_v}$$

where:

- $\sum P$ is the electrical power used by the fan (or sum of all fans in the ventilation system) [kW]
- q_v is the gross amount of air circulated through the fan (or ventilation system) [m³/s].

There are various sub-definitions of SFP for different specific applications, including SFP_e (building energy performance calculations), SFP_v (for performance verification tests), SFP_i (individual fan), SFP_{AHU} (air handling unit), SFP_{FCU} (fan coil unit), and SFP_{BLDG} (whole building). These are explained in and in part in. Reference 1 also describes how account for intermittently operated fans, e.g. kitchen hoods, and part-load performance in variable air volume (VAV) systems.

SFP can be expressed in the following equivalent SI units:

$$[\mathrm{SFP}] \equiv \frac{\mathrm{kW}}{\mathrm{m^3/s}} \equiv \frac{\mathrm{W}}{\mathrm{l/s}} \equiv \frac{\mathrm{kJ}}{\mathrm{m^3}} \equiv \mathrm{kPa}$$

SFP and Fan System Efficiency

As you can see above, SFP can be expressed in units of pressure, since pressure is a measure of energy per m^3 air. The relationship between SFP, fan pressure rise, and fan system efficiency is simply:

$$\eta_{tot} \cdot \mathrm{SFP} = \Delta p_{tot}$$

where:

- η_{tot} is the overal efficiency of the driven fan system [-]
- Δp_{tot} is the total pressure rise across the fan [kPa]

In the case of an ideal lossless fan system (i.e. $\eta_{tot} = 1$) the SFP is exactly equal to the fan pressure rise (i.e. total pressure loss in the ventilation system). In reality the fan system efficiency is often in the range 0 to 60% (i.e. $\eta_{tot} < 0.6$); it is lowest for small fans or inefficient operating points (e.g. throttled flow or free-flow). The efficiency is a function of the total losses in the fan system, including aerodynamic losses in the fan, friction losses in the drive (e.g. belt), losses in the electric motor, and variable speed drive power electronics. For more insight into how to maximise energy efficiency and minimise noise in fan systems.

Whole-house Fan

A whole-house fan is a type of fan, or exhaust system commonly venting into a building's attic, designed to pull hot air out of the building. It is sometimes confused with an attic fan.

A whole-house fan pulls hot air out of a building and forces it into the attic space. This causes a positive pressure in the attic forcing air out through the gable and/or soffit vents, while at the same time producing a negative pressure inside the living areas which draws cool air in through open windows.

Attic fans, by comparison, only serve to remove some hot air from the attic; no direct cooling effect is provided to the actual living space.

History

Whole-house fans were mainly popularised in the Southern United States through the 1950s-60s, as they were much cheaper and easier to find than air conditioners and still cooled relatively well.

Benefits

A whole-house fan can significantly lower the temperature in a building very quickly, and is much less expensive to operate than air conditioning. Newer whole house fans can be environmentally friendly and energy efficient additions to house cooling systems. On temperate days they can be turned on to circulate rising hot air out of the house while pulling cool air in. Also new models are quieter and smaller than their older counterparts.

Drawbacks

A major issue to watch for is backdrafts from open flames like water heaters - If the WHF creates a lot of negative pressure it can cause unsealed furnaces to spew toxic carbon-monoxide into the house. Consult an expert HVAC person to address this.

Air is drawn into the house through open windows, meaning that the air is unfiltered and may contain pollen or other allergens, which settle out on floors and furniture making the house get dirty fast. This is in contrast to an air conditioner, which mainly circulates air through a heat exchanger.

The fan can also be noisy. The noise is due to the large volume of air handled by such a fan and the speed of the fan blades. Direct drive motors spin the blades faster, producing more noise compared to belt-driven fans, which allow lower fan speeds and less noise. The noise may interfere with other normal household activities such as listening to music, watching television, or conversation.

If not of an insulated design, the fan serves as a large "hole in the ceiling" through which heat will escape during the winter time.

Types

There are two types of fan:

- Ceiling Mounted: Mounted on ceiling between the attic and living space.
- Ducted: Remotely mounted away from the ceiling; can exhaust heat from multiple locations; operation is extremely quiet.

Installation

Attics must be properly vented so that the air is able to escape back outdoors as quickly as it is being blown into the attic. A good rule of thumb is to provide the same amount of venting as the size of the shutter hole in the ceiling.

How Does A Whole House Fan Save Energy

The following graphic depicts the temperature trends of the ambient air (outside), a house using a whole house fan, and a house just using air conditioning. I gathered the data over several different days and chose the ones that best represent a typical summer day in Southwestern Oregon. The timeline starts at nightfall and transitions to daytime.

Both houses are trying to maintain a 75 degree temperature or less, so you'll notice that when either house reaches 75 degrees, the air conditioning is turned on.

Below the houses are circles that represent the relative amount of energy used in each house to maintain the temperature at 75 degrees or less over the course of a night/day. From my estimates, using a whole house fan in combination with central air uses 500 times less energy than solely relying on central air to cool your house. I will be adding more details to this page as time goes on.

Saving Money with a Whole House Fan

Energy Usage Comparison

A typical central air conditioner uses at least 2000 watts per hour, or 2kWh, and this may be on the smallish side.

A newer, quiet and efficient whole house fan that moves air at the rate of 1530 cubic feet per minute uses 44 watts per hour, or 0.044kWh.

2000kWh/44kWh=45.45, so a whole house fan uses 45 times less energy than a central air conditioner using the above numbers.

Therefore, the energy circle associated with the central air conditioning grows at a rate 45 times the amount of the energy circle associated with the whole house fan. The circles grow at the same rate when both houses are using air conditioning.

Given electricity rates reported by the Department of Energy of ~11 cents/kWh, the above example would cost $4.40 per day if using only central air conditioning. When using a whole house fan, this is reduced to $0.45 per day. If this occurs 60 times during the summer, you've saved over $235.00 in one summer.

Electricity rates are increasing according to these official Department of Energy statistics, more than doubling over the past 20 years. If we extrapolate these rate increases, then the savings 10 years from now would accumulate to over 3000 dollars, paying for the whole house fan and then some.

Evaporative Cooler

An evaporative cooler (also swamp cooler, desert cooler, and wet air cooler) is a device that cools air through the evaporation of water. Evaporative cooling differs from typical air conditioning systems which use vapour-compression or absorption refrigeration cycles. Evaporative cooling works by employing water's large enthalpy of vapourisation. The temperature of dry air can be dropped significantly through the phase transition of liquid water to water vapour, which requires much less energy than refrigeration. In extremely dry climates, it also has the added benefit of conditioning the air with more moisture for the comfort of occupants. Unlike refrigeration, it requires a water source, and must continually consume water to operate.

In the United States, the use of the term *swamp cooler* may be due to the odor of algae produced by early units Air washers and wet cooling towers use the same principles as evaporative coolers but are optimised for purposes other than cooling the air inside a building. For example, an evaporative cooler may be designed to cool coils of a large air conditioning system to increase its efficiency.

Evaporative cooling is especially well suited for climates where the air is hot and humidity is low. In the United States, the western/ mountain states are good locations, with evaporative coolers prevalent in cities like Denver, Salt Lake City, Albuquerque, El Paso, Tucson, and Fresno where sufficient water is available. Evaporative air conditioning is also popular and well-suited to the southern (temperate) part of Australia. In dry, arid climates, the installation and operating cost of an evaporative cooler can be much lower than that of refrigerative air conditioning, often by 80% or so. However, evaporative cooling and vapour-compression air conditioning are sometimes used in combination to yield optimal cooling results. Some evaporative coolers may also serve as humidifiers in the heating season.

In locations with moderate humidity there are many cost-effective uses for evaporative cooling, in addition to their widespread use in dry climates. For example, industrial plants, commercial kitchens, laundries, dry cleaners, greenhouses, spot cooling (loading docks, warehouses, factories, construction sites, athletic events, workshops, garages, and kennels) and confinement farming (poultry ranches, hog, and dairy) often employ evaporative cooling. In highly humid climates, evaporative cooling may have little thermal comfort benefit beyond the increased ventilation and air movement it provides.

History

The evaporative cooler was the subject of numerous U.S. patents in the twentieth century; many of these, starting in 1906, suggested or assumed the use of excelsior (wood wool) pads as the elements to bring a large volume of water in contact with moving air to allow evaporation to occur. A typical design, as shown in a 1945 patent, includes a water reservoir (usually with level controlled by a float valve), a pump to circulate water over the excelsior pads and a squirrel-cage fan to draw air through the pads and into the house. This design and this material remain dominant in evaporative coolers in the American Southwest, where they are also used to increase humidity.

Evaporative cooling was in vogue for aircraft engines in the 1930s, for example with the Beardmore Tornado airship engine. Here the system was used to reduce, or eliminate completely, the radiator which would otherwise create considerable drag. In these systems the water in the engine was kept under pressure with pumps, allowing it to heat to temperatures above 100°C, as the actual boiling point is a function of the pressure. The superheated water was then sprayed through a nozzle into an open tube, where it flashed into steam, releasing its heat. The tubes could be placed under the skin of the aircraft, resulting in a zero-drag cooling system. However these systems also had serious disadvantages. Since the amount of tubing needed to cool the water was large, the cooling system covered a significant portion of the plane even though it was hidden. This added complexity and reliability issues. In addition this large size meant it was very easy for it to be hit by enemy fire, and practically impossible to armour. British and U.S. developers used ethylene glycol instead, cooling the liquid in radiators. The Germans instead used streamlining and positioning of traditional radiators. Even the method's most ardent supporters, Heinkel's Günter brothers, eventually gave up on it in 1940.

Externally-mounted evaporative cooling devices to cool interior air were used in some automobiles, often as aftermarket accessories, until modern vapour-compression air conditioning became widely available.

Physical Principles

Evaporative cooling is a physical phenomenon in which evaporation of a liquid, typically into surrounding air, cools an object or a liquid in contact with it. Latent heat, the amount of heat that is needed to evaporate the liquid, is drawn from the air. When considering water evaporating into air, the wet-bulb temperature, as compared to the air's dry-bulb temperature, is a measure of the potential for evaporative cooling.

The greater the difference between the two temperatures, the greater the evaporative cooling effect. When the temperatures are the same, no net evaporation of water in air occurs, thus there is no cooling effect.

A simple example of natural evaporative cooling is perspiration, or sweat, which the body secretes in order to cool itself. The amount of heat transfer depends on the evaporation rate, however for each kilogram of water vapourised 2257 kJ of energy (about 890 BTU per pound of pure water, at 95°F) are transferred. The evaporation rate in turn depends on the humidity of the air and its temperature, which is why one's sweat accumulates more on hot, humid days: the perspiration cannot evaporate.

Evaporative cooling is not the same principle as that used by vapour-compression refrigeration units, although that process also requires evaporation (although the evaporation is contained within the system).

In a vapour-compression cycle, after the refrigerant evaporates inside the evaporator coils, the refrigerant gas is compressed and cooled, causing it to return to its liquid state. In contrast an evaporative cooler's water is only evaporated once. In a space-cooling unit the evaporated water is introduced into the space along with the now-cooled air; in an evaporative tower the evaporated water is carried off in the airflow.

Applications

Evaporative cooling is a common form of cooling buildings for thermal comfort since it is relatively cheap and requires less energy than other forms of cooling. However, evaporative cooling requires an abundant water source as an evaporate, and is only efficient when the relative humidity is low, restricting its effective use to dry climates. Evaporative cooling also raises the internal humidity level significantly, which can cause problems such as lumpy salt, swelling of wood panelling, doors and trim, pianos going out of tune or suffering internal rusting, etc.

Evaporative cooling is commonly used in cryogenic applications. The vapour above a reservoir of cryogenic liquid is pumped away, and the liquid continuously evaporates as long as the liquid's vapour pressure is significant. Evaporative cooling of ordinary helium forms a 1-K pot, which can cool to at least 1.2 K. Evaporative cooling of helium-3 can provide temperatures below 300 mK. These techniques can be used to make cryocoolers, or as components of lower-temperature cryostats such as dilution refrigerators. As the temperature decreases, the vapour pressure of the liquid also falls, and cooling becomes less effective. This sets a lower limit to the temperature attainable with a given liquid.

An application of a similar principle to evaporative cooling is the "self-refrigerating" beverage can. A separate compartment inside the can contains a desiccant and a liquid. Just before drinking a tab is pulled so that the desiccant comes into contact with the liquid and dissolves. As it does so it absorbs an amount of heat energy called the latent heat of fusion. Evaporative cooling works with the phase change of liquid into vapour and the latent heat of vapourisation but the self-cooling can uses the change from solid to liquid and the latent heat of fusion to achieve the same result.

On Earth, trees transpire large amounts of water through pores in their leaves called stomata, and through this process of evaporative cooling, forests interact with climate at local and global scales.

Evaporative cooling has been observed to operate on a planetary scale on Pluto, where it has been called an anti-greenhouse effect.

Evaporative cooling is also the last cooling step in order to reach the ultra-low temperatures required for Bose–Einstein condensation (BEC). Here, so-called forced evaporative cooling is used to selectively remove high-energetic ("hot") atoms from an atom cloud until the remaining cloud is cooled below the BEC transition temperature. For a cloud of 1 million alkali atoms, this temperature is about 1μK.

Evaporative Cooler Designs

All designs take advantage of the fact that water has one of the highest known enthalpy of vapourisation (latent heat of vapourisation) values of any common substance.

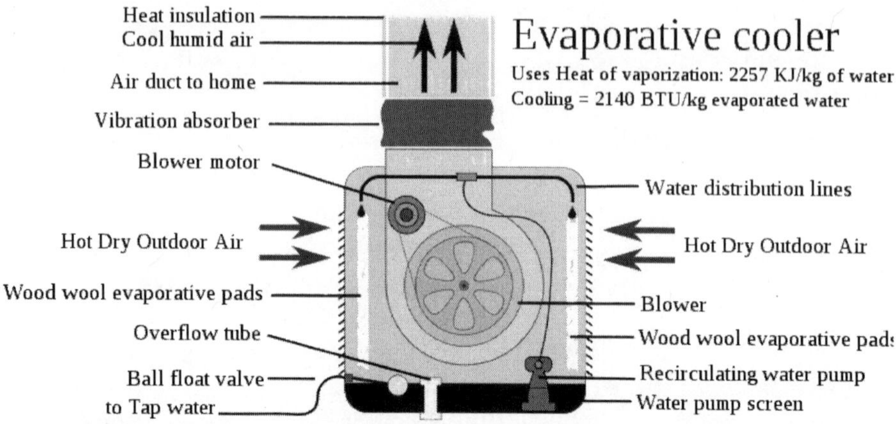

Direct evaporative cooling (open circuit) is used to lower the temperature of air by using latent heat of evaporation, changing liquid water to water vapour. In this process, the energy in the air does not change. Warm dry air is changed to cool moist air. The heat of the outside air is used to evaporate water.

Indirect evaporative cooling (closed circuit) is similar to direct evaporative cooling, but uses some type of heat exchanger. The cooled moist air never comes in direct contact with the conditioned environment.

Two-stage evaporative cooling, or indirect-direct. Traditional evaporative coolers use only a fraction of the energy of vapour-compression or absorption air conditioning systems.

Unfortunately, except in very dry climates they increase humidity to a level that makes occupants uncomfortable. Two-stage evaporative coolers do not produce humidity levels as high as that produced by traditional single-stage evaporative coolers.

In the first stage of a two-stage cooler, warm air is pre-cooled indirectly without adding humidity (by passing inside a heat exchanger that is cooled by evaporation on the outside). In the direct stage, the pre-cooled air passes through a water-soaked pad and picks up humidity as it cools.

Since the air supply is pre-cooled in the first stage, less humidity is needed in the direct stage to reach the desired cooling temperatures. The result, according to manufacturers, is cooler air with a relative humidity between 50 and 70 percent, depending on the climate, compared to a traditional system that produces about 70–80 percent relative humidity air.

Typical

Typically, residential and industrial evaporative coolers use direct evaporation and can be described as an enclosed metal or plastic box with vented sides containing a centrifugal fan or 'blower', electric motor with pulleys (known as 'sheaves' in HVAC), (or a direct-driven axial fan), and a water pump to wet the evaporative cooling pads. The units can be mounted on the roof (down draft, or downflow), or exterior walls or windows (side draft, or horizontal flow) of buildings. To cool, the fan draws ambient air through vents on the unit's sides and through the damp pads. Heat in the air evaporates water from the pads which are constantly re-dampened to continue the cooling process. Thus cooled, moist air is then delivered to the building via a vent in the roof or wall.

Because the cooling air originates outside the building, one or more large vents must exist to allow air to move from inside to outside. Air should only be allowed to pass once through the system, or the cooling effect will decrease. This is due to the air reaching the saturation point. Often 15 or so air changes per hour (ACHs) occur in spaces served by evaporative coolers.

Cooler Pads

Traditionally, evaporative cooler pads consist of excelsior (wood wool) (aspen wood fibre) inside a containment net, but more modern materials, such as some plastics and melamine paper, are entering use as cooler-pad media. Wood absorbs some of the water, which allows the wood fibres to cool passing air to a lower temperature than some synthetic materials.

Evaporative (wet) Cooling Towers

Cooling towers are structures for cooling water or other working media to near-ambient wet-bulb temperature. Wet cooling towers operate on the evaporative cooling principle, but are optimised to cool the water rather than the air. Cooling towers can often be found on large buildings or on industrial sites. They transfer heat to the environment from chillers, industrial processes, or the Rankine power cycle, for example.

Misting Systems

Misting systems work by forcing water via a high pressure pump and tubing through a brass and stainless steel mist nozzle that has an orifice of about 5 micrometres, thereby producing a micro-fine mist. The water droplets that create the mist are so small that they instantly flash evaporate. Flash evaporation can reduce the surrounding air temperature by as much as 35 F° (20 C°) in just seconds. For patio systems, it is ideal to mount the mist line approximately 8 to 10 feet (2.4 to 3.0 m) above the ground for optimum cooling. Misting is used for applications such as flowerbeds, pets, livestock, kennels, insect control, odor control, zoos, veterinary clinics, produce cooling and greenhouses.

Figure: Mist spraying system with water pump beneath

Misting fans

A misting fan is similar to a humidifier. A fan blows a fine mist of water into the air. If the air is not too humid, the water evaporates, absorbing heat from the air, allowing the misting fan to work as an air conditioner. A misting fan may be used outdoors, especially in a dry climate.

Performance

Understanding evaporative cooling performance requires an understanding of psychrometrics. Evaporative cooling performance is dynamic due to changes in external temperature and humidity level. A residential cooler should cool air to within 3–4 C° (5–7 F°) of the corresponding wet-bulb temperature.

It is simple to predict cooler performance from standard weather report information. Because weather reports usually contain the dewpoint and relative humidity, but not the wet-bulb temperature, a Psychrometric chart must be used to identify the wet bulb temperature. Once the wet bulb temperature and the dry bulb temperature are identified the cooling performance or leaving air temperature of the cooler may be determined:

$$T_{LA} = T_{DB} - ((T_{DB} - T_{WB}) \text{ x } E)$$

T_{LA} = Leaving Air Temp

T_{DB} = Dry Bulb Temp

T_{WB} = Wet Bulb Temp

E = Efficiency of the evaporative media.

Evaporative media efficiency usually runs between 80% to 90% and the evaporation efficiency drops very little over time. Typical aspen pads used in residential evaporative coolers offer around 85% efficiency while CELdek type of evaporative media offer efficiencies of 90% + depending on air velocity. The celdec media is more often used in large commercial and industrial installations.

As an example, in Las Vegas, Nevada, with a typical summer design day of 108°F DB/66°F WB or about 8% relative humidity, the leaving air temperature of a residential cooler would be:

T_{LA} = 108° – ((108° – 66°) x 85% efficiency)

T_{LA} = 72.3°F

However, either of two methods can be used to estimate performance:

Use a Psychrometric chart to calculate wet bulb temperature, and then add 6–8 F° as described above.

Use a rule of thumb which estimates that the wet bulb temperature is approximately equal to the ambient temperature, minus one third of the difference between the ambient temperature and the dew point. As before, add 6–8 F° as described above.

Some examples clarify this relationship:

At 32 °C (90 °F) and 15% relative humidity, air may be cooled to nearly16 °C (61 °F). The dew point for these conditions is 2 °C (36 °F).

At 32 °C (90 °F) and 50% relative humidity, air may be cooled to about 24 °C (75 °F). The dew point for these conditions is 20 °C (68 °F).

At 40 °C (104 °F) and 15% relative humidity, air may be cooled to nearly 21 °C (70 °F). The dew point for these conditions is 8 °C (46 °F).(*Cooling examples extracted from the June 25, 2000 University of Idaho publication, "Homewise"*).

Because evaporative coolers perform best in dry conditions, they are widely used and most effective in arid, desert regions such as the southwestern USA and northern Mexico.

The same equation indicates why evaporative coolers are of limited use in highly humid environments: for example, a hot August day in Tokyo may be 30 °C (86 °F), 85% relative humidity, 1,005 hPa pressure. This gives dew point 27.2 °C (81.0 °F) and wet-bulb temperature 27.88 °C (82.18 °F). According to the formula above, at 85% efficiency air may be cooled only down to 28.2 °C (82.8 °F) which makes it quite impractical.

Comparison to Air Conditioning

Less Expensive to Install

Estimated cost for installation is about half that of central refrigerated air conditioning.

Less Expensive to Operate

Estimated cost of operation is 1/4 that of refrigerated air.

Power consumption is limited to the fan and water pump vs. compressors, pumps and blowers.

Ease of Maintenance

The only two mechanical parts in most basic evaporative coolers are the fan motor and the water pump, both of which can be repaired at low cost and often by a mechanically inclined homeowner.

Ventilation Air

The constant and high volumetric flow rate of air through the building reduces the age-of-air in the building dramatically.

Evaporative cooling increases humidity, which, in dry climates, may improve the breathability of the air.

The pad itself acts as a rather effective air filter when properly maintained; it is capable of removing a variety of contaminants in air, including urban ozone caused by pollution, regardless of very dry weather. Refrigeration-based cooling systems lose this ability whenever there is not enough humidity in the air to keep the evaporator wet while providing a constant trickle of condensate that washes out dissolved impurities removed from the air.

Performance

High dewpoint (humidity) conditions decrease the cooling capability of the evaporative cooler. No dehumidification. Traditional air conditioners remove moisture from the air, except in very dry locations where recirculation can lead to a buildup of humidity. Evaporative cooling adds moisture, and in dry climates, dryness may improve thermal comfort at higher temperatures.

Comfort

The air supplied by the evaporative cooler is typically 80–90% relative humidity; very humid air reduces the evaporation rate of moisture from the skin, nose, lungs, and eyes.

High humidity in air accelerates corrosion, particularly in the presence of dust. This can considerably shorten the life of electronic and other equipment.

High humidity in air may cause condensation. This can be a problem for some situations (e.g., electrical equipment, computers, paper/books, old wood).

Water

Evaporative coolers require a constant supply of water to wet the pads.

Water high in mineral content will leave mineral deposits on the pads and interior of the cooler. Bleed-off and refill (purge pump) systems may reduce this problem. These deposits will include radioactive isotopes if the water is from a radioactive source such as deep wells in certain Texas aquifers. Such deposits would tend to concentrate the radioactivity, whether by crystallisation of dissolved radioactive elements onto the pads or accumulation of radioactive particulates on the pads.

The water supply line may need protection against freeze bursting during off-season, winter temperatures. The cooler itself needs to be drained too, as well as cleaned periodically and the pads replaced.

Computer Fan

A computer fan is any fan inside, or attached to, a computer case used for cooling purposes, and may refer to fans that draw cooler air into the case from the outside, expel warm air from inside, or move air across a heatsink to cool a particular component. The use of fans to cool a computer is an example of active cooling

Usage of a Cooling Fan

As processors, graphics cards, RAM and other components in computers have increased in speed and power consumption, the amount of heat produced by these components as a side-effect of normal operation has also increased. These components need to be kept within a specified temperature range to prevent overheating, instability, malfunction and damage leading to a shortened component lifespan.

While in earlier personal computers it was possible to cool most components using natural convection (passive cooling), many modern components require more effective active cooling. To cool these components, fans are used to move heated air away from the components and draw cooler air over them. Fans attached to components are usually used in combination with a heatsink to increase the area of heated surface in contact with the air, thereby improving the efficiency of cooling.

In the IBM compatible PC market, the computer's power supply unit (PSU) almost always uses an exhaust fan to expel warm air from the PSU. Active cooling on CPUs started to appear on the Intel 80486, and by 1997 was standard on all desktop processors. Chassis or case fans, usually one exhaust fan to expel heated air from the rear and optionally an intake fan to draw cooler air in through the front, became common with the arrival of the Pentium 4 in late 2000. A third vent fan in the side of the PC, often located over the CPU, is also common. The graphics processing unit (GPU) on many modern graphics cards also requires a heatsink and fan. In some cases, the northbridge chip on the motherboard has another fan and heatsink. Other components such as the hard drives and RAM may also be actively cooled, though as of 2007 this remains relatively unusual. It is not uncommon to find five or more fans in a modern PC.

Chapter 4

Cooling Fan Application

Figure: An 80x80x25mm computer fan

Case Mount

Used to aerate the case of the computer. The components inside the case cannot dissipate heat efficiently if the surrounding air is too hot. Case fans move air through the case, usually drawing cooler outside air in through the front (where it may also be drawn over the internal hard drive racks) and expelling it through the rear. There may be a third fan in the side or top of the case to draw outside air into the vicinity of the CPU, which is usually the largest single heat source. Standard case fans are 80 mm, 92 mm or 120 mm in width and length. As case fans are often the most readily visible form of cooling on a PC, decorative fans are widely available and may be lit with LEDs, made of UV-reactive plastic, and covered with decorative grilles. Decorative fans and accessories are popular with case modders. Air filters are often used over intake fans, to prevent dust from entering the case.

A power supply (PSU) fan often plays a double role, not only keeping the PSU itself from overheating, but also removing warm air from inside the case. PSUs with two fans are also available, which typically have a fan on the inside to supply case air into the PSU and a second fan on the back to expel the heated air.

CPU Fan

Used to cool the CPU (central processing unit) heatsink. Effective cooling of a concentrated heat source such as a large-scale integrated circuit requires a heatsink, which may be cooled by a fan; use of a fan alone will not prevent overheating of the small chip.

Graphics Card Fan

Used to cool the heatsink of the graphics processing unit or the memory on graphics cards. These fans were not necessary on older cards because of their low power dissipation, but most modern graphics cards, especially those designed for 3D graphics and gaming, need their own dedicated cooling fans. Some of the higher powered cards can produce more heat than the CPU (dissipating up to 289 watts), so effective cooling is especially important. Passive coolers for new video cards, however, are not unheard of, such as the Thermalright HR-03.

Chipset Fan

Used to cool the heatsink of the northbridge of a motherboard's chipset; this may be needed where the system bus is significantly overclocked and dissipates more power than is usual, but may otherwise be unnecessary.

Other Types of Fans

Other less commonly encountered fans may include:

- PCI slot fan: A fan mounted in one of the PCI slots, usually to supply additional cooling to the PCI and/or graphics cards.
- Hard disk fan: A fan mounted next to or on a hard disk drive. This may be desirable on faster-spinning (e.g. 10,000 RPM) hard disks with greater heat production.
- CD burner fan: Some internal CD and/or DVD burners included cooling fans.

Physical Characteristics

Square-framed fans are usually used, but round frames are also used, often so that a larger fan than the mounting holes would otherwise allow can be used (e.g., a 120 mm fan with holes for the corners of a 90 mm square fan). The width of square fans and the diameter of round ones are usually stated in millimeters; common sizes include 40, 60, 80, 92, and 120 mm. Heights are typically 10 or 20 mm, but this is usually not an important dimension as it does not affect mounting holes or apertures in the case. The airflow that fans generate is typically stated in cubic feet per minute (CFM), and the speed of rotation in revolutions per minute (RPM). Where noise is an issue large, slower-turning fans are quieter, and sensors and circuits can be used that reduce fan speed, and hence noise, when temperature is not high.

The other consideration when choosing a computer fan is static pressure. A fan with high static pressure is more effective at forcing air through restricted spaces, such as the gaps between a radiator or heatsink; static pressure is more important than CFM when choosing a fan for use with a heatsink. The relative importance of static pressure depends on the degree to which the airflow is restricted by geometry; static pressure becomes more important as the spacing between heatsink fins decreases. Static pressure is usually stated in either mm Hg or mm H_2O.

The type of bearing used in a fan can affect its performance and noise output. Most computer fans use one of the following bearing types: Sleeve bearings use two surfaces lubricated with oil or grease as a friction contact. They often use porous sintered sleeves to be self-lubricating, requiring only infrequent maintenance or replacement. Sleeve bearings are less durable at higher temperatures as the contact surfaces wear and the lubricant dries up, eventually leading to failure; however, lifetime is similar at relatively low ambient temperatures Sleeve bearings may be more likely to fail at higher temperatures, and may perform poorly when mounted in any orientation other than vertical. The lifespan of a sleeve-bearing fan may be around 40,000 hours at 50 °C. Fans that use sleeve bearings are generally cheaper than fans that use ball bearings, and are quieter at lower speeds early in their life, but can become noisy as they age.

Rifle bearings are similar to sleeve bearings, but are quieter and have almost as much lifespan as ball bearings. The bearing has a spiral groove in it that pumps fluid from a reservoir. This allows them to be safely mounted horizontally (unlike sleeve bearings), since the fluid being pumped lubricates the top of the shaft. The pumping also ensures sufficient lubricant on the shaft, reducing noise, and increasing lifespan.

Ball bearings: Though generally more expensive, ball bearing fans do not suffer the same orientation limitations as sleeve bearing fans, are more durable at higher temperatures, and are quieter than sleeve-bearing fans at higher rotation speeds. The lifespan of a ball bearing fan may be over 60,000 hours at 50 °C.

Fluid bearings have the advantages of near-silent operation and high life expectancy (comparable to ball bearings), but tend to be the most expensive. Magnetic bearings or maglev bearings, in which the fan is repelled from the bearing by magnetism.

Fan Sizing

Fans are available in many sizes and capacities. Typically 120mm case and power supply fans are used where cooling requirements are demanding, as for computers used to play games, and for quieter operation at lower speeds; cheaper 80mm fans are used for less demanding cases. In general, the faster the fan, the noisier it is. Within a given physical size capacity is roughly proportional to current draw for a given flow a larger fan will be quieter than a smaller fan.

Fan Connector

Connectors usually used for computer fans (there is no agreed standard) are:

3-pin Molex connector KK Family

This connector is used when connecting a fan to the motherboard or other circuit board. It is a small thick rectangular in-line female connector with two tabs on the outer-most edge of one long side. The size and spacing of the pin sockets is identical to a standard 3-pin female IC connector. The three pins are used for ground, +12 V power, and a tachometer signal. Molex Part number of receptacle is 22-01-3037. Molex Part number of individual crimp contacts is 08-55-0101. Matching PCB header Molex 22-23-2031

4-pin Molex connector KK Family

This is a special variant of the Molex KK connector with four pins but with the locking/polarisation features of a 3-pin connector. The additional pin is used for a pulse-width modulation signal to provide variable speed control. These can be plugged into 3-pin headers, but will lose their fan speed control. Molex Part number of receptacle is 47054-1000. Molex Part number of individual crimp contacts is 08-55-0101. Molex Part number of the header is 47053-1000.

4-pin Molex connector

This connector is used when connecting the fan directly to the power supply. It consists of two wires (red/12V and black/ground) leading to and splicing into a large in-line 4-pin male-to-female Molex connector. This is the same connector as used on hard drives before the SATA became standard.

Dell proprietary

This connector is an expansion of a simple 3-pin female IC connector by adding two tabs to the middle of the connector on one side and a lock-tab on the other side. The size and spacing of the pin sockets is identical to a standard 3-pin female IC connector and 3-pin Molex connector. Some models have the wiring of the white wire (speed sensor) in the middle, whereas the standard 3-pin Molex requires the white wire as pin #3, thus compatibility issues may exist.

Alternatives

If a fan is not desirable, because of noise, reliability, or environmental concerns, there are some alternatives. Some improvement can be achieved by eliminating all fans except one in the power supply which also draws hot air out of the case.

Sinking motherboards in liquid oil provides excellent convection cooling and protects from humidity and water without the need for heatsinks or fans. Special care must be taken to ensure compatibility with adhesives and sealants used on the motherboard and ICs. This solution is used in some external environments such as wireless equipment located in the wild.

Fan Control

Fan control is the management of the rotational speed of an electric fan. There are many types of electric fans and many types of fan controls. This article principally describes the control of a computer fan, to provide adequate cooling while lowering noise.

Need for fan control

As modern PCs grow more powerful so do their requirements for electrical power. Computers convert most of this electrical power into heat generated by all major components.

Some early generation PCs did not need active ventilation. Power supplies eventually needed forced cooling, and soon took up the duty of cooling the rest of the PC with the ATX standard. The byproduct of increased heat generation is that the fan(s) need to move increasing amounts air and thus, need to be more powerful. Since they must move more air through the same area of space, fans will naturally become more noisy.

In fact, if one installs extra fans in a PC case, the noise levels can reach 70 dB. Since fan noise increases with the fifth power of the fan rotation speed, reducing rotations per minute (RPM) by a small amount potentially means a reduction in fan noise. This must be done cautiously, as excessive reduction in speed may cause components to overheat and be damaged. If done properly fan noise can be drastically reduced.

Fan types

Common cooling fans have at least two pins on the connector. These two pins connect to the fan motor, usually a brushless DC type. Extra pins may be present, with either or both of the following features:

- A tachometer that measures the actual speed of the fan as a pulse train, frequency being proportional to speed.
- Control of fan speed using PWM

Two-pin fans operate either as an on/off fan, or can be controlled by varying the voltage.

Fans will have either two, three, or four pins on the connector.

Types of Control

No Control

The simplest method of fan control is simply to leave the fan on all the time. This creates quite a bit of noise and increases power requirements, but keeps the system the coolest.

Thermostatic

In this style of fan control, the fan is either on or off. A system thermistor checks the temperature inside the chassis, and if it detects a temperature outside of range, it spins the fans up to maximum. When the temperature drops below a threshold again, the fans are turned back off. This control method reduces power requirements during periods of low usage, but when the system is operating at capacity, the fan noise can become a problem again.

Linear voltage regulation

A standard cooling fan is essentially a bladed DC motor. By varying the voltage input across the acceptable range for a fan, the speed of the fan will increase (to added voltage) and decrease (to reduced voltage). A faster fan, means more air moved and thus a higher heat exchange rate. There are a few ways to perform this regulation.

Resistors

Resistors are the simplest method of reducing fan noise, but they add to the heat generated inside the computer case. Since the voltage drop is proportional to the current, the fan may not start. They need to be of the appropriate power rating. For variable fan control, potentiometers could be used along with a transistor such as a MOSFET whose output voltage is controlled by the potentiometer. It is possible to use a rheostat instead.

Diodes

A diode in series with the fan will reduce the voltage being output to the fan. You can use a silicon diode with a fixed voltage drop of about 0.7V per diode. The power rating should be noted.

Voltmodding

Voltmodding describes the practice of varying the voltage fed to a component; in this case, a computer fan. This can be achieved by connecting the ground wire of the fan to the +5V rail and the positive wire to the +12V rail of a typical PC power supply to achieve a theoretical +7V (positive seven volts). However, this is a potentially risky method: the parts on the +5V power line might be exposed to overvoltage in case of a short in the fan. A less common variation is to increase the voltage to the fan by connecting the ground wire to a -12V rail (located on the motherboard ATX connector) and the positive wire to the +12V rail, producing a total of +24V24V can destroy the fans, other options are 17V and 15.3V.

This is not a good idea, if there is nothing else on +5V rail. Because it will cause the current to go into +5V instead out of it as normal. PSU is not designed for that.

Pulse-width modulation

Pulse-width modulation (PWM) is a common method of controlling computer fans. A PWM capable fan is usually connected to a 4-pin connector (pinout: Ground, +12V, sense, control). The sense pin is used to measure the rotation speed of the fan and the control pin is a open-drain or open-collector output, which requires a pull-up to 5V or 3.3V in the fan. Unlike linear voltage regulation, where the fan voltage is proportional to the speed, the fan is driven with a constant supply voltage; the speed control is performed by the fan based on the control signal.

The control signal is a square wave operating at 25kHz, with the duty cycle determining the fan speed. Typically a fan can be driven between about 30% and 100% of the rated fan speed, using a signal with up to 100% duty cycle. The exact speed behaviour (linear, off until a threshold value, or a minimum speed until a threshold) at low control levels is manufacturer-dependent.

Fan speed controllers

Figure: A fan controller with LEDs to indicate fan status and potentiometer and switch to control fan speed.

Another method, popular with gamers, is the manual fan speed controller. They can be mounted in an expansion slot, a 5.25" or 3.5" drive bay or come built into the computer's case. Using switches or knobs, attached fans can have their speeds adjusted by one of the above methods.

Software

The method by which the software physically controls the fan is usually PWM. Many companies now provide software to control fan speeds on their motherboards under Microsoft Windows.

- AOpen motherboards can use "SilentTEK".
- ASUS motherboards can use "Q-Fan".
- MSI motherboards can use "Core Centre".
- Universal abit motherboards can use "µGuru".
- Gigabyte motherboards can use "EasyTune 6".

- Intel desktop boards (older socket 478 etc.) use "Active Monitor" and "Desktop Control Centre".
- Intel desktop boards (newer socket 775 etc.) use "Desktop Utilities".
- Dell Inspiron/Latitude/Precision computers can use "I8kfanGUI"
- Various computers can use the freeware "SpeedFan".

Computers running Linux can use lm sensors.

Computer Cooling

Computer cooling is required to remove the waste heat produced by computer components, to keep components within their safe operating temperature limits. Various cooling methods help to improve processor performance or reduce the noise of cooling fans.

Components which produce heat and are susceptible to performance loss and damage include integrated circuits such as CPUs, chipset and graphics cards, along with hard drives (though excessive cooling of hard drives has been found to have negative effects). Overheated parts fail early and may give sporadic problems resulting in system freezes or crashes.

Both integral and peripheral means are used to keep the temperature of each component at a safe level. With regard to integral means, CPU and GPUs are designed with energy efficiency, including heat dissipation, in mind; though improved efficiency may only allow increased performance instead of reduced heat. Peripheral means include heat sinks to increase the surface area which dissipates heat, fans to speed up the exchange of air heated by the computer parts for cooler ambient air, and in some cases softcooling, the throttling of computer parts in order to decrease heat generation.

As a safety measure, many computers are designed to turn themselves off if the internal temperature exceeds a certain point. Alternatively, some have an option in their BIOS that allows the user to determine if the system emits an alarm beep or shuts itself down when the core temperature reaches the level set by the user. However, setting this incorrectly can result in hardware damage or erratic system behaviour

Causes of Heat Build Up

The amount of heat generated by an integrated circuit (e.g., a

CPU or GPU), the prime cause of heat build up in modern computers, is a function of the efficiency of its design, the technology used in its construction and the frequency and voltage at which it operates.

Figure: The dust buildup on this laptop CPU heat sink after three years of use has made the laptop unusable due to frequent thermal shutdowns.

In operation, the temperature of a computer's components will rise until the heat lost to the surroundings is equal to the heat produced by the component, and thus the temperature of the component reaches equilibrium. For reliable operation, the equilibrium temperature must be sufficiently low for the structure of the computer's circuits to survive.

Cooling can be hindered by:

- Dust acting as a thermal insulator and impeding airflow, thereby reducing heat sink and fan performance.
- Poor airflow including turbulence due to friction against impeding components such as ribbon cables, or improper orientation of fans, can reduce the amount of air flowing through a case and even create localised whirlpools of hot air in the case.
- Poor heat transfer due to a lack of, or poor application of thermal compounds and sufficient surface area of heat sinks to radiate off the heat.

Damage prevention

Thermal sensors in some CPUs and GPUs can shut down the computer when high temperatures are detected. However, reliance on such measures may not prevent repeated incidents from permanently damaging the integrated circuit.

An integrated circuit may also shut down parts of the circuit when it is idling, or to scale back the clock speed under low workloads or high temperatures, with the goal of reducing both power use and heat generation.

Air cooling

Fans are most commonly used for air cooling. A *computer fan* may be attached to the computer case, or attached to a CPU, GPU, chipset, PSU, hard drive or PCI slot. Common fan sizes include 40, 60, 80, 92, 120, and 140 mm. Recently, 200mm fans have begun to creep into the performance market, as well as even larger sizes such as 230 and 240mm.

In desktops

Figure: Typical airflow through a desktop ATX case.

Desktop computers typically use one or more fans for cooling. Almost all desktop power supplies have at least one fan to exhaust air from the case. Most manufacturers recommend bringing cool, fresh air in at the bottom front of the case, and exhausting warm air from the top rear.

If there is more air being forced into the system than is being pumped out (due to an imbalance in the number or strength of fans), this is referred to as a "positive" airflow, as the pressure inside the unit would be higher than outside. A balanced or neutral airflow is the most efficient although a slightly positive airflow results in less dust build up if dust filters are used. Negative pressure inside the case can create problems such as clogged optical drives due to sucking in air (and dust).

In high density computing

Data centres typically contain many racks of flat 1U servers. Air is drawn in at the front of the rack and exhausted at the rear. Because data centres typically contain such large numbers of computers and other power-consuming devices, they risk overheating of the various components if no additional measures are taken. Thus, extensive HVAC systems are used. Often a raised floor is used so the area under the floor may be used as a large plenum for cooled air and power cabling.

Another way of accommodating large numbers of systems in a small space are blade chassis. In contrast to the horizontal orientation of flat servers, blade chassis are often oriented vertically. This vertical orientation facilitates convection. When the air is heated by the hot components, it tends to flow to the top on its own, creating a natural air flow along the boards. This stack effect can help to achieve the desired air flow and cooling. Some manufacturers expressly take advantage of this effect.

In Laptop Computing

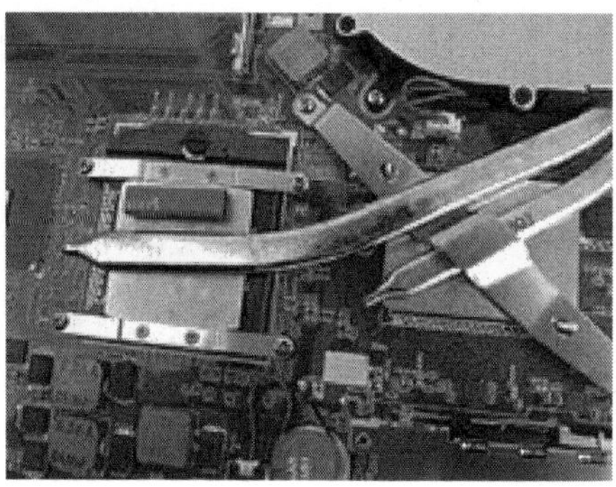

Figure: Laptops use separated apparatus for heatsinks, heat is drawn from the processor using copper pipes and an exhaust fan expels it from the device. Pictured here are the contact points on a CPU and a GPU of a Sony Vaio E series laptop.

Figure: The heat is expelled from a laptop using a remote exhaust

Most laptops use air cooling in order to keep the CPU and other components within their operating temperature range. Because the fan's air is forced through a small port, the fan and heatsinks can be clogged by dust or be obstructed by objects placed near the port. This can cause overheating, and can be a cause of component failure in laptops. The severity of this problem varies with laptop design, its use and power dissipation. With recent reductions in CPU power dissipation, this problem can be anticipated to reduce in severity.

Liquid Submersion Cooling

An uncommon practice is to submerge the computer's components in a thermally conductive liquid. Personal computers that are cooled in this manner do not generally require any fans or pumps, and may be cooled exclusively by passive heat exchange between the computer's parts, the cooling fluid and the ambient air. Extreme component density supercomputers such as the Cray-2 and Cray T90 used additional liquid to chilled liquid heat exchangers in order to facilitate heat removal.

The liquid used must have sufficiently low electrical conductivity in order for it not to interfere with the normal operation of the computer's components. If the liquid is somewhat electrically conductive, it may be necessary to insulate certain parts of components susceptible to electromagnetic interference, such as the CPU. For these reasons, it is preferred that the liquid be dielectric.

Liquids commonly used in this manner include various liquids invented and manufactured for this purpose by 3M, such as Fluorinert. Various oils, including but not limited to cooking, motor and silicone oils have all been successfully used for cooling personal computers.

Evaporation can pose a problem, and the liquid may require either to be regularly refilled or sealed inside the computer's enclosure. Liquid may also slowly seep into and damage components, particularly capacitors, causing an initially functional computer to fail after hours or days immersed.

Waste Heat Reduction

Where full-power, full-featured modern computers are not required, some companies opt to use less powerful computers or computers with fewer features. For example: in an office setting, the IT department may choose a thin client or a diskless workstation thus cutting out the heat-laden components such as hard drives and optical disks. These devices are also often powered with direct current from an external power supply brick which still wastes heat, but not inside the computer itself.

The components used can greatly affect the power consumption and hence waste heat. A VIA EPIA motherboard with CPU typically generates approximately 25 watts of heat whereas a Pentium 4 motherboard typically generates around 140 watts. While the former has considerably less computing power, both types are adequate and responsive for tasks such as word processing and spreadsheets. Choosing a LCD monitor rather than a CRT can also reduce power consumption and excess room heat, as well as the added benefit of increasing available physical desk space.

Conductive and Radiative Cooling

Some laptop components, such as hard drives and optical drives, are commonly cooled by having them make contact with the computer's frame, increasing the surface area which can radiate and otherwise exchange heat. Several new 2GB and above DDR2 and DDR3 sticks of RAM, or system memory, are conductive cooled via a finned heatsink that is clipped onto the top edge of the memory stick. This also holds true for video cards that use a finned heatsink over the GPU for silent PC applications.

Spot Cooling

In addition to system cooling, various individual components usually have their own cooling systems in place. Components which are frequently individually cooled include, but are not limited to, the CPU, GPU and the Northbridge chip. Some cooling solutions employ one or more methods of cooling, and may also utilise logic and/or temperature sensors in order to vary the power used in active cooling components.

Passive Heat-sink Cooling

Passive heat-sink cooling involves attaching a block of machined or extruded metal to the part that needs cooling. A thermal adhesive may be used. More commonly for a personal-computer CPU, a clamp holds the heat sink directly over the chip, with a thermal grease or thermal pad spread between. This block usually has fins and ridges to increase its surface area. The heat conductivity of metal is much better than that of air, and it radiates heat better than the component that it is protecting (usually an integrated circuit or CPU). Until recently, fan-cooled aluminium heat sinks were the norm for desktop computers. Today, many heat sinks feature copper base-plates or are entirely made of copper.

Figure: Passive heatsink on an Intel GMA graphics chip.

Dust buildup between the metal fins of a heat sink gradually reduces efficiency, but can be countered with a gas duster by blowing away the dust along with any other unwanted excess material.

Passive heat sinks are commonly found on older CPUs, parts that do not get very hot (such as the chipset), and low-power computers.

Usually a heat-sink is attached to the integrated heat spreader (IHS), essentially a large, flat plate attached to the CPU, with conduction paste layered between. This dissipates or spreads the heat locally. Unlike a heat sink, a spreader is meant to redistribute heat, not to remove it. In addition, the IHS protects the fragile CPU.

Passive cooling involves no fan noise.

Active heat-sink cooling

Active heat-sink cooling uses the same principle as passive, with the addition of a fan that blows over or through the heat sink. The air movement increases the rate at which the heat sink can exchange heat with the ambient air. Active heat sinks are the primary method of cooling modern processors and graphics cards.

Figure: Active heat sink with a fan and heat pipes.

The buildup of dust is greatly increased with active heat-sink cooling, because the fan continually takes in the dust present in the surrounding air.

Peltier Cooling or Thermoelectric Cooling

In 1821 T. J. Seebeck discovered that different metals, connected at two different junctions, will develop a micro-voltage if the two junctions are held at different temperatures. This effect is known as the "Seebeck effect"; it is the basic theory behind the TEC (thermoelectric cooling).

In 1834 Jean Peltier discovered the inverse of the Seebeck effect, now known as the "Peltier effect". He found that applying a voltage to a thermocouple creates a temperature differential between two sides. This results in an effective, albeit extremely inefficient heat pump.

Modern TECs use several stacked units each composed of dozens or hundreds of thermocouples laid out next to each other, which allows for a substantial amount of heat transfer. A combination of bismuth and tellurium is most commonly used for thermocouples.

As active heat pumps, TECs can cool the surface of components below ambient temperatures. This is impossible with common radiator cooled water cooling systems and heatpipe HSFs.

Water Cooling

While originally limited to mainframe computers, water cooling has become a practice largely associated with overclocking in the form of either manufactured kits, or in the form of do-it-yourself setups assembled from individually gathered parts. The past few years has seen water cooling increasing its popularity with pre-assembled, moderate to high performance, desktop computers. Water has the ability to dissipate more heat from the parts being cooled than the various types of metals used in heatsinks, making it suitable for overclocking and high performance computer applications.

Figure: DIY Water cooling setup showing 12v pump, CPU Waterblock and the typical application of a T-Line

Advantages to water cooling include the fact that a system is not limited to cooling one component, but can be set up to cool the central processing unit, graphics processing unit, and/or other components at the same time with the same system. As opposed to air cooling, water cooling is also influenced less by the ambient temperature. Water cooling's comparatively low noise-level compares favourably to that of active cooling, which can become quite noisy. One disadvantage to water cooling is the potential for a coolant leak. Leaked coolant can damage any electronic components with which it comes into contact. Another drawback to water cooling is the complexity of the system; an active heat sink is much simpler to build, install, and maintain than a water cooling solution.

The IBM Aquasar system uses *hot water cooling* to achieve energy efficiency, the water being used to heat buildings as well.

Heat Pipe

A heat pipe is a hollow tube containing a heat transfer liquid. As the liquid evaporates, it carries heat to the cool end, where it condenses and then returns to the hot end (under capillary action, or, in earlier implementations, under gravitation). Heat pipes thus have a much higher effective thermal conductivity than solid materials. For use in computers, the heat sink on the CPU is attached to a larger radiator heat sink. Both heat sinks are hollow as is the attachment between them, creating one large heat pipe that transfers heat from the CPU to the radiator, which is then cooled using some conventional method. This method is expensive and usually used when space is tight (as in small form-factor PCs and laptops), or absolute quiet is needed (such as in computers used in audio production studios during live recording). Because of the efficiency of this method of cooling, many desktop CPUs and GPUs, as well as high end chipsets, use heat pipes in addition to active fan-based cooling to remain within safe operating temperatures.

Phase-change Cooling

Phase-change cooling is an extremely effective way to cool the processor. A vapour compression phase-change cooler is a unit which usually sits underneath the PC, with a tube leading to the processor. Inside the unit is a compressor of the same type as in a window air conditioner. The compressor compresses a gas (or mixture of gases) which condenses it into a liquid. Then, the liquid is pumped up to the processor, where it passes through an expansion device, this can be from a simple capillary tube to a more elaborate thermal expansion valve. The liquid evaporates (changing phase), absorbing the heat from the processor as it draws extra energy from its environment to accommodate this change. The evaporation can produce temperatures reaching around "15 to "150 degrees Celsius. The gas flows down to the compressor and the cycle begins over again. This way, the processor can be cooled to temperatures ranging from "15 to "150 degrees Celsius, depending on the load, wattage of the processor, the refrigeration system and the gas mixture used. This type of system suffers from a number of issues but, mainly, one must be concerned with dew point and the proper insulation of all sub-ambient surfaces that must be done (the pipes will sweat, dripping water on sensitive electronics).

Alternately, a new breed of cooling system is being developed, inserting a pump into the thermo siphon loop. This adds another degree of flexibility for the design engineer, as the heat can now be effectively transported away from the heat source and either reclaimed or dissipated to ambient. Junction temperature can be tuned by adjusting the system pressure; higher pressure equals higher fluid saturation temperatures. This allows for smaller condensers, smaller fans, and/or the effective dissipation of heat in a high ambient environment. These systems are, in essence, the next generation liquid cooling paradigm, as they are approximately 10 times more efficient than single phase water. Since the system uses a dielectric as the heat transport media, leaks do not cause a catastrophic failure of the electric system.

This type of cooling is seen as a more extreme way to cool components, since the units are relatively expensive compared to the average desktop. They also generate a significant amount of noise, since they are essentially refrigerators, however the compressor choice and air cooling system is the main determinant of this, allowing for flexibility for noise reduction based on the parts chosen.

Liquid Nitrogen

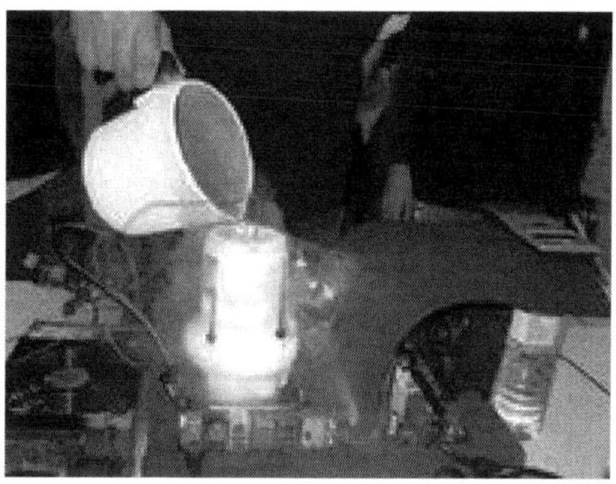

Figure: Liquid nitrogen may be used to cool an overclocked PC.

As liquid nitrogen boils at -196 °C, far below the freezing point of water, it is valuable as an extreme coolant for short overclocking sessions.

In a typical installation of liquid nitrogen cooling, a copper or aluminium pipe is mounted on top of the processor or graphics card. After being heavily insulated against condensation, the liquid nitrogen is poured into the pipe, resulting in temperatures well below -100 °C.

Evaporation devices ranging from cut out heat sinks with pipes attached to custom milled copper containers are used to hold the nitrogen as well as to prevent large temperature changes. However, after the nitrogen evaporates, it has to be refilled. In the realm of personal computers, this method of cooling is seldom used in contexts other than overclocking trial-runs and record-setting attempts, as the CPU will usually expire within a relatively short period of time due to temperature stress caused by changes in internal temperature. Although liquid nitrogen is non-flammable, it can condense oxygen directly from air. Mixtures

Liquid Helium

Liquid helium, colder than liquid nitrogen, has also been used for cooling. Liquid helium evaporates at -269 °C, and temperatures ranging from -230 to -240 °C have been measured from the heatsin. But Helium is more rare and more expensive than Nitrogen. Also, extremely low temperatures can cause integrated circuits to stop functioning.

Soft Cooling

Softcooling is the practice of utilising software to take advantage of CPU power saving technologies to minimise energy use. This is done using halt instructions to turn off or put in standby state CPU subparts that aren't being used or by underclocking the CPU.

Undervolting

Undervolting is a practice of running the CPU or any other component with voltages below the device specifications. An undervolted component draws less power and thus produces less heat. The ability to do this varies by manufacturer, product line, and even different production runs of the same exact product (as well as that of other components in the system), but modern processors are typically shipped with voltages higher than strictly necessary. This provides a buffer zone so that the processor will have a higher chance of performing correctly under sub-optimal conditions, such as a lower quality mainboard (motherboard). However, too low a voltage will not allow the processor to function correctly, producing errors, system freezes or crashes, or the inability to turn the system on. Undervolting too far does not typically lead to hardware damage, though in worst-case scenarios, program or system files can be corrupted.

This technique was generally employed by those seeking low-noise systems, as less cooling is needed because of the reduction of heat production. Since the popularity of hand-held or remote computers (Unmanned vehicles, mobile & cordless phones/camera/viewers, etc), undervolting is used to prolong battery endurance.

Integrated Chip Cooling Techniques

Conventional cooling techniques all attach their "cooling" component to the outside of the computer chip, or via IHS and/or heat sinks. This "attaching" technique will always exhibit some thermal resistance, reducing its effectiveness. The heat can be more efficiently and quickly removed by directly cooling the local hot spots. At these locations, power dissipation of over 300W/cm^2 (typical CPU are less than 100W/cm^2, although future systems are expected to exceed 1000W/cm^2 can occur. This form of local cooling is essential to developing high power density chips. This ideology has led to the investigation of integrating cooling elements into the computer chip. Currently there are two techniques: micro-channel heat sinks, and jet impingement cooling.

In micro-channel heat sinks, channels are fabricated into the silicon chip (CPU), and coolant is pumped through them. The channels are designed with very large surface area which results in large heat transfers. Heat dissipation of 3000W/cm^2 has been reported with this technique. In comparison to the Sun power density of around 7400W/cm^2. The heat dissipation can be further increased if two-phase flow cooling is applied. Unfortunately the system requires large pressure drops, due to the small channels, and the heat flux is lower with dielectric coolants used in electronic cooling. Another local chip cooling technique is jet impingement cooling. In this technique, a coolant is flown through a small orifice to form a jet. The jet is directed toward the surface of the CPU chip, and can effectively remove large heat fluxes. Heat dissipation of over 1000W/cm^2 has been reported. The system can be operated at lower pressure in comparison to the micro-channel method. The heat transfer can be further increased using two-phase flow cooling and by integrating return flow channels (hybrid between micro-channel heat sinks and jet impingement cooling).

Cooling and Overclocking

Extra cooling is usually required by those who run parts of their computer (such as the CPU and GPU) at higher voltages and frequencies than manufacturer specifications call for, called overclocking. Increasing performance by this modification of settings results in a greater amount of heat generated and thus increasing the risk of damage to components and/or premature failure.

The installation of higher performance, non-stock cooling may also be considered modding. Many overclockers simply buy more efficient, and often, more expensive fan and heat sink combinations, while others resort to more exotic ways of computer cooling, such as liquid cooling, Peltier effect heatpumps, heat pipe or phase change cooling. There are also some related practices that have a positive impact in reducing system temperatures:

Heat Sink Lapping

Figure: Milled surface of a processor cooler

Heat sink lapping is the smoothing and polishing of the contact (bottom) part of a heat sink to increase its heat transfer efficiency. The desired result is a contact area which has a more even surface, as a less even contact surface creates a larger amount of insulating air between the heat sink and the computer part it is attached to. Polishing the surface using a combination of fine sandpaper and abrasive polishing liquids can produce a mirror-like shine, an indicator of a very smooth metal surface. Even a curved surface can become extremely reflective, yet not particularly flat, as is the case with curved mirrors; thus heat sink quality is based on *overall flatness*, more than optical properties. Lapping a high quality heat sink can damage it, because, although the heat sink may become shiny, it is likely that more material will be removed from the edges, making the heat sink less effective overall.

If attempted, a piece of float glass should be used, as it self-levels as it cools and offers the most economical solution to producing a perfectly flat surface.

Use of Exotic Thermal Conductive Compounds

Some overclockers use special thermal compounds whose manufacturers claim to have a much higher efficiency than stock thermal pads. Heat sinks clean of any grease or other thermal transfer compounds have a very thin layer of these products applied, and then are placed normally over the CPU. Many of these compounds have a high proportion of silver as their main ingredient due to its high thermal conductivity. The resulting difference in the temperature of the CPU is measurable (several celsius degrees), so the heat transfer does appear to be superior to stock compounds. Some people experience negligible gains and have called to question the advantages of these exotic compounds, calling the style of application more important than the compound itself. Also note that there may be a 'setting' or 'curing' period and negligible gains may improve over time as the compound reaches its optimum thermal conductivity.

Use of Rounded Cables

Most older PCs use flat ribbon cables to connect storage drives (IDE or SCSI). These large flat cables greatly impede airflow by causing drag and turbulence. Overclockers and modders often replace these with rounded cables, with the conductive wires bunched together tightly to reduce surface area. Theoretically, the parallel strands of conductors in a ribbon cable serve to reduce crosstalk (signal carrying conductors inducing signals in nearby conductors), but there is no empirical evidence of rounding cables reducing performance. This may be because the length of the cable is short enough so that the effect of crosstalk is negligible. Problems usually arise when the cable is not electromagnetically protected and the length is considerable, a more frequent occurrence with older network cables.

These computer cables can then be cable tied to the chassis or other cables to further increase airflow.

This is less of a problem with new computers that use Serial ATA which has a much narrower cable.

Airflow Optimisation

The colder the cooling medium (the air), the more effective the cooling. Cooling air temperature can be improved with these guidelines:

- Supply cool air to the hot components as directly as possible. Examples are air snorkels and tunnels that feed outside air directly and exclusively to the CPU or GPU cooler. For example, the BTX case design prescribes a CPU air tunnel.
- Expel warm air as directly as possible. Examples are: Conventional PC (ATX) power supplies blow the warm air out the back of the case. Many dual-slot graphics card designs blow the warm air through the cover of the adjacent slot. There are also some aftermarket coolers that do this. Some CPU cooling designs blow the warm air directly towards the back of the case, where it can be ejected by a case fan.

- Air that has already been used to spot-cool a component should not be reused to spot-cool a different component (this follows from the previous items). The ATX case design can be said to violate this rule, since the power supply gets its "cool" air from the inside of the case, where it has been warmed up already. The BTX case design also violates this rule, since it uses the CPU cooler's exhaust to cool the chipset and often the graphics card.
- Prefer cool intake air, avoid inhaling exhaust air (outside air above or near the exhausts). For example, a CPU cooling air duct at the back of a tower case would inhale warm air from a graphics card exhaust. Moving all exhausts to one side of the case, conventionally the back, helps to keep the intake air cool.
- Hiding cables behind motherboard tray or simply apply ziptie and tucking cables away to provide unhindered airflow.

Fewer fans strategically placed will improve the airflow internally within the PC and thus lower the overall internal case temperature in relation to ambient conditions. The use of larger fans also improves efficiency and lowers the amount of waste heat along with the amount of noise generated by the fans while in operation.

There is little agreement on the effectiveness of different fan placement configurations, and little in the way of systematic testing has been done. For a rectangular PC (ATX) case, a fan in the front with a fan in the rear and one in the top has been found to be a suitable configuration. However, AMD's (somewhat outdated) system cooling guidelines notes that "A front cooling fan does not seem to be essential. In fact, in some extreme situations, testing showed these fans to be recirculating hot air rather than introducing cool air." It may be that fans in the side panels could have a similar detrimental effect—possibly through disrupting the normal air flow through the case. However, this is unconfirmed and probably varies with the configuration

Quiet PC

History

Prior to about 1975, all computers were typically large industrial/commercial machines, often in a centralised location with a dedicated room-sized cooling system. For these systems noise was not an important issue.

With the development of the home computer, early systems such as the Commodore 64 were very low wattage and were often fanless. If there was a fan, it was a low-speed fan only used to cool the power supply, such as in the IBM PC XT.

Fan noise only started to become an issue as CPU processing power increased. Processors up to about 60 megahertz did not require anything more than a single case fan and a passive heatsink. Beyond that point, a fan would be installed over the CPU heatsink to blow air straight down onto the processor, in what is known as *spot-cooling*. There was no regard for where the intake air came from, or where exhaust was going. The sole purpose of the fan was to move heat from a small concentrated location under the heatsink into the larger air mass inside the computer case.

As desktop computers grew in performance, more fans were included to provide spot-cooling in many more specific locations where heat dissipation was needed, without regard to overall airflow or trying to do thermal analysis of cooling efficiency.

- Originally, video display controllers were fairly low power devices without any need for active cooling. But with the development of the 3D graphics card, it became common for the video card to have its own fan, separate from a general case/system fan. The development of 3D cards working in tandem required separate spot-cooling fans for each card.
- Multiprocessor systems such as the Pentium Pro typically needed a separate spot-cooling fan for each CPU.

Computer cases often have not been designed to consider the overall airflow of the system, while spot-cooling fans only focus on cooling a specific location without regard to where the exhaust air is going. Sometimes fan airflow is not coordinated, such as with the power supply and case fans both blowing air in or sucking air out, with no other venting. This combination could lead to a system with a large number of internal spot-cooling fans that is overheating because there is poor overall airflow into and out of the case.

The Green PC

With the developing societal interest in energy conservation, the concept of designing systems to only consume as much power as is needed at a particular moment has helped to reduce both power consumption and system noise.

Initially this was not important for large industrial/commercial systems and for home computers, and systems typically operated at full power consumption all the time, with cooling systems also designed to operate at maximum cooling capacity all the time. The screensaver for example, is an animated motion of a computer screen designed to prevent image burn-in, used at a time when computer monitors tended to stay powered on all the time, or only turned off by operation of a physical switch.

Energy conservation first started to become an issue with the development of portable computers that had limited battery power; reducing power usage directly translated into longer operation time. System noise is also a factor for laptops since the noise producing components cannot be shifted away to another location. The first low power and energy conserving CPUs were developed for use in laptops, while desktops continued to operate at fixed high power levels.

Large industrial systems such as network file servers and database servers were the last to get power conservation measures added, since their primary focus is high performance centralised operation. However, as system density and power usage increased, businesses began to recognise that server systems do not need to run at full power all the time, and there is money to be saved using a server that reduces power consumption when it is not needed. Individually the cost savings is small but for a large business with hundreds to thousands of systems, the savings can be significant.

Causes of noise

The main causes of PC noise are:

- Mechanical friction noise generated by micro motors and fan bearings, as well as vibration noise from low quality chassis and improper assemblies.
- Turbulence caused by obstructions in the flow of air, such as poorly designed fan grilles and heatsinks. A lack of clearance between rotating fan blades and nearby support struts and grilles will create an audible noise, similar to how a mechanical siren operates. As the fan blades spin, they produce a vortex of air that trails off the edge of the blade. When the vortexes path crosses an object it may produce noise. Fan blades can be designed to reduce this problem by using special notched shapes.

- Noise generated by electrical coils or transformers used in power supplies, motherboards, video cards or LCD monitors.

Noise in personal computers has been increasing with rising computing power and number of transistors on a single die (integrated circuit). More transistors of a given size use more power, which releases more heat. Faster-rotating cooling fans are one common way to remove this heat. Also, motor rotation speed for hard disc drive (HDD) and Optical disc drive (ODD) has been rising for faster data processing with technical advances in micro motors. Faster rotation (usually measured in RPM) causes higher bearing friction, thus more noise, given the same bearing technology. The noise issue had received widespread attention with AMD's early Athlon CPUs and Intel's Pentium 4 Prescott core CPU known for its excessive heat and bundled fan noise running on high RPM. With the introduction of Home Theatre PCs (HTPC), the excessive heat and noise problem, that had been mostly confined to the overclocking and quiet computing communities, came to the attention of the general public.

The main approaches to reducing noise problems from personal computers are:

1. Reduce heat generation by using energy efficient parts - nearly all the energy used by a computer is converted into heat.
2. Improve cooling - by using more efficient cooling parts and lower friction, quieter bearings.
3. Use Soundproofing to reduce the effects of remaining noise sources.

Noise Reduction Methods

Figure: Reducing noise with new CPU cooler.

Figure: This passive heat sink in a Powermac G4 relies on large surface areas

Common noise reduction methods

- Replace heat sinks with more efficient models. This often entails the use of larger copper or aluminium heat sinks which may incorporate heat pipes.
- Replace fans with passive cooling solutions where possible, such as fans on motherboards and GPUs.
- Replace fans with low-speed, large-diameter fans with low bearing and motor noise. Larger fans can move more air per revolution than smaller fans.
- Replace the constant-speed fan in the power supply with a thermostatically controlled variable speed fan that runs at less than maximum speed, and thus runs quieter most of the time.
- Replace the power supply with a quieter model. The main considerations, from a noise-reduction point of view, in choosing a power supply are fan quality, AC/DC conversion efficiency, and how good the thermal fan speed control is at keeping the fan running slow and steady. Efficiency is important because the less heat that is produced the less work the fan has to perform.
- Replace hard drives with quieter models. Hard drives can also be replaced with laptop hard drives, with solid state devices like compact flash or networked file systems like NFS. This reduces or eliminates this source of noise and reduces system power and cooling requirements.
- Place a damping material around hard drives (or other spinning drives) such as Sorbothane
- Cover the case with sound insulation material such as rubber, foam or fiber mat, although this method has limited

effectiveness. The material can (because of its weight) dampen case resonance, and can also absorb some high-frequency sound. Care must be taken to be sure the soundproofing does not interfere with airflow and cooling.

- In energy-hungry computers, water cooling may be necessary for quiet operation. Older water pumps sometimes can make systems noisier than air-cooled, low-power computers. However, recent advances in 12v DC pump technologies have resulted in highly reduced noise levels from many pumps. In a modern watercooling system geared towards silence rather than performance, the loudest component in the computer is often the Hard Drive or Optical disc drive when performing accesses to the media.

Low-cost methods

A number of methods exist for reducing computer noise at little or no added cost.

- Reduce CPU supply voltage ("undervolting"). Many of today's CPUs can run stably at their stock speed, or even with a slight overclock, at a reduced voltage, which reduces heat output. Underclocking can be done for the same effect, however this reduces performance and is not as effective as undervolting; all the same, underclocking may allow further undervolting. Power consumption is approximately proportional to $V^2 f$, that is, it varies linearly with the clock frequency and quadratically with the voltage. This means that even a small reduction in voltage can have a large effect in power consumption. Undervolting and underclocking can also be used with chipsets and GPUs.
- Enable Cool'n'Quiet for AMD CPUs or EIST on Intel's CPUs.
- Reduce fan speed. For newer computers, the speed of fans can be varied automatically, depending on how hot certain parts of the computer get. Lowering a DC fan motor's supply voltage will reduce its speed while making it quieter and lowering the amount of air the fan moves. Doing this arbitrarily could lead to components overheating; therefore, whenever performing hardware work it is advised to monitor the temperature of system components. Fans with Molex connectors can be modified easily. With 3-pin fans, either fixed inline resistors or diodes, or commercial fan controllers, such as the Zalman Fanmate,

can be used. Software like speedfan may allow fan speed control. Many newer motherboards support Pulse-width modulation (PWM) control, allowing the fan speed to be set in the BIOS or with software.

- Mount fans on anti-vibration mounts.
- Remove restrictive fan grills to allow easier airflow, or replace noisy fan grills with quieter versions.
- Use software such as Nero DriveSpeed to reduce the speed of optical drives.
- Isolate hard disk noise, either by using anti-vibration mounts (generally rubber or silicone grommets), or by suspending the hard disk to fully decouple it from the computer chassis by mounting it in a 5.25 inch drive bay with viscoelastic polymer mounts.
- Set the hard disk's AAM value to its lowest setting. This reduces the seek noise produced by the hard drive, but also reduces performance slightly.
- Set operating system to spin down hard drives after a short time of inactivity. This may reduce a drive's life span and commonly conflicts with the OS and running programs, though it can still be useful for drives that are only used for data storage.
- Defragment the hard drives to reduce the drive heads' need to search widely for data, also improves performance.
- Arrange components and cables to improve airflow. Wires hanging inside the computer can block the airflow, which can increase the temperature. They can be easily moved to the side of the case so that air can pass through more easily
- Remove dust from inside the computer. Dust on computer parts will retain more heat. Fans draw in dust along with outside air, it can build up quickly inside the computer. Dust can be removed with a vacuum cleaner, gas duster or compressed air. Special anti-static vacuum cleaners should be used however to prevent electrostatic discharge (ESD). Ideally, this would be done often enough to prevent a significant amount of dust from ever building up. How frequently this would need to be performed would depend entirely on the environment in which the computer is used.

Sound Power and Pressure Measurement

Though standards do exist for measuring and reporting sound power output by such things as computer components, they are often ignored. Many manufacturers do not give sound power measurements. Some report sound pressure measurements, but those that do often do not specify how sound pressure measurements were taken. Even such basic information as measurement distance is rarely reported. Without knowing how it was measured, it is not possible to verify these claims, and comparisons between such measurements (e.g. for product selection) are meaningless. Comparative reviews, which test several devices under the same conditions are more useful. Although even then, an average sound pressure level is only one factor in determining which components will be perceived as quieter

Individual Components in a Quiet PC

The following are notes regarding individual components in quiet PCs.

The motherboard, CPU, and video card are major energy users in a computer. Components that need less power will be easier to cool quietly. A quiet power supply is selected to be efficient while providing enough power for the computer.

Motherboards

Figure: Passively cooled northbridge chipsets help reduce noise.

A motherboard based on a chipset that uses less energy may be easier to cool quietly.

Many modern motherboard chipsets have hot northbridges (notably nForce4), which may come with active cooling, usually a small, noisy fan. Fanless heatsinks, such as the Zalman ZM-NB47J, ZM-NBF47 or the Thermalright HR-05, may be used to eliminate a noisy chipset fan. Some motherboard manufacturers have replaced these fans by incorporating large heatsinks or heatpipe coolers, however they still require good case airflow to remove heat. Also, motherboard voltage regulators often have heatsinks and may need airflow to ensure adequate cooling.

Motherboards can also produce coil noise.

Undervolting and underclocking generally require motherboard support.

Some motherboards can control the fan speed using software like speedfan. Most recent motherboards have built in PWM based fan control for one or two fans.

CPUs

The heat output of a CPU can vary according to its brand and model - to be exact, its TDP. Intel's third revision Pentium 4, using the "Prescott" core, was infamous for being one of the hottest-running CPUs on the market. By comparison, AMD's Athlon series and the new Intel Core 2 perform better at lower clock speeds, and thus produce less heat.

Modern CPUs often incorporatc energy saving systems, such as Cool'n'Quiet, LongHaul, and SpeedStep. These reduce the CPU clock speed and core voltage when the processor is idle, thus reducing heat. The heat produced by CPUs can be further reduced by undervolting, underclocking or both.

Most modern mainstream and value CPUs are made with a lower TDP to reduce heat, noise, and power consumption. Most of Intel's desktop Core 2 Duo processors have 65W TDPs, and AMD has newer processors with a TDP between 35W and 65W. Some processors come in special low power versions.

Maximum TDP:

- AMD Fusion C and E-series: 9W-18W
- Athlon 64 X2: 45W, 65W, 85W
- Intel Core and Core 2 series: 35W, 65W, 85W
- Intel Pentium M and Celeron M (lacks SpeedStep)
- Intel Xeon L55xx and L56xx series: 60W
- VIA C7: 12W-20W (Fanless)
- transmeta processors

- Geode (processor): 5W, 25W
- Intel Atom: 2W, 4W, 8W
- Loongson: 1W, 5W, 7W (Fanless).

Video Cards

Video cards can produce a significant amount of heat. A fast GPU may be the largest power consumer in a computer. For instance, 161 watts peak power consumption for an ATI Radeon HD 2900 XT 512. Because of space limitations, video card coolers often use small fans running at high speeds, making them noisy.

Display options for making a quiet computer include:

- Use motherboard video output - typically motherboard video takes less power than an external video card, typically at the price of lower gaming or HD video decoding performance.
- Select a video card that does not use a fan
  - o many of the more efficient GPUs are available in fanless models
  - o many older video cards used less power than more recent video cards, thus heatsinks or fans were often not required
- Replace the GPU cooler with a larger heatsink and possibly a larger, slower fan
- No video card - use a terminal, thin client, USB display or desktop sharing software if display required

Power Supplies

PSUs are made quieter through the use of higher efficiency (which reduces waste heat and need for airflow), quieter fans, more intelligent fan controllers (ones for which the correlation between temperature and fan speed is more complex than linear), more effective heatsinks and through designs which allow air to flow through with less resistance.

For a given power supply size, more efficient supplies, such as those certified 80 plus, generate less heat.

Selecting a power supply of appropriate wattage for the computer is important for high efficiency and minimising heat. Power supplies are typically less efficient when lightly or heavily loaded. High wattage power supplies will typically be less efficient when lightly loaded, for instance when the computer is idle or sleeping. Most desktop computers spend most of their time lightly loaded For example, most desktop PCs draw less than 250 watts at full load, and 200 watts or less is more typical.

Power supplies with thermally controlled fans can be made quieter by providing a cooler and/or less obstructed source of air. For instance, the power supply is in a separate compartment in the Antec P180 to keep the air supplied to the PSU cool.

The fan in a power supply can be replaced with a quieter one, although there is a risk of electric shock when doing this, and it usually voids the warranty.

Fanless power supplies are available.

- Some of them are equipped with large passive heat sinks and rely on convection or case airflow to dissipate heat. It is also imperative that such fanless power supplies be installed in a case with good ventilation.
- There are also fanless DC to DC power supplies that operate like those in laptops, using an external power brick to supply DC power, which is then converted to appropriate voltages and regulated for use by the computer. These power supplies usually have lower wattage ratings.

The electrical coils in power supplies can produce noise which can become noticeable in a quiet PC.

Cases

Figure: Antec P180, with isolated chambers for more segregated airflow.

Figure: Another example of the Antec P180, this one demonstrating the use of the Scythe Ninja, a fanless CPU cooler.

Cases designed for low noise usually include reasonably quiet fans, and often come with a relatively quiet power supply. Some cases for quiet computers incorporate heatsinks to cool components passively.

Cases that provide lots of space make it easier to quiet a PC, both by allowing for airflow and by accommodating large coolers.

Case airflow

Noise optimised cases like the Antec P180 and Antec P150 often have ducting and partitioning within the case to optimise airflow and thermally isolate components. For example, the P180 has the PSU mounted in the bottom of the case in an isolated partition. This design feature allows cooler air to enter the PSU, reducing the necessary airflow and accordingly, the noise output of the fan. Apple has also employed this tactic in their G5 workstations in an effort to reduce noise. Antec's *Sonata* is often considered by the mainstream to be one of the quietest PC cases; however, it has since been surpassed by the P180 and other more-advanced cases. Vents and ducts may easily be added to regular cases. More obstructive fan grills increase pressure drop and lower airflow, necessitating higher fan speeds and more noise output. They also increase the turbulence of the flow, which causes some noise of its own. Cases designed to be quiet typically have wire grills or honeycombed fan grills, which perform almost as well as wire grills; both are far superior to the old style of stamped grill.

Features that facilitate neat cable management, such as brackets and space to run cables behind motherboard tray, help increase cooling efficiency.

Air filtering

Air filters can help to prevent dust from coating heatsinks and surfaces, thereby impeding heat transfer, making fans spin faster. However the filter itself can increase noise if it restricts airflow too much, or is not cleaned, requiring a larger or faster fan to handle the pressure drop behind the filter. Regular cleaning of air intake filters will help to keep airflow minimally restricted and keep the interior clean.

In some cases a fine mesh intake screen is sufficient to stop most large dust particles from entering the system. These screens should be vacuumed or washed to remove dust.

Case Soundproofing

The inside of a case can be lined with dampening materials to reduce noise by:

- attenuating the vibration of the case panels via extensional damping or constrained-layer damping.
- reducing the amplitude of the vibration of the case panels by increasing their mass.
- absorbing airborne noise, such as with foam.

Cooling Systems

Heat sinks that operate efficiently with little airflow are often used in quiet computers. Typically they are (relatively) large, and have larger spaces to allow freer airflow. Often heat pipes are used to help distribute heat. For instance, in 2007, the Scythe Ninja or the Thermalright Ultra-120 were frequently used as CPU heat sinks in quiet computers.

Fans

Figure: A 120 mm variable speed fan.

Bearing and motor noise vary between different fan models and often between different samples of the same model.

Quiet PCs typically use larger (e.g. 120 mm) low-speed fans. Although 140 mm fans are made by some manufacturers, such as Aerocool and Yate Loon, there are very few cases or heatsinks that can use them. Fan adapters, which allow larger fans to be used in place of smaller ones, and fan brackets, like the Zalman FB123, often help when replacing small fans.

Quiet fan manufacturers include Nexus, EBM-Papst, Yate Loon and Scythe. In situations where the resistance to flow is very low, like in free-air conditions, Noctua fans also perform very well. Extensive comparative surveys have been posted by Silent PC Review and MadShrimps.

Fan controllers can be used to slow down fans and to precisely choose fan speed. Fan controllers can produce a fixed fan speed using an inline resistor or diode, or a variable speed using a potentiometer or Pulse Width Modulation (PWM). Resistor-based fan control feeds the fan a lower voltage, while PWM fan control rapidly cycles between feeding the fan full voltage and no voltage. PWM fan control reduces rotational speed, and is the easiest and most efficient option for motherboards which have PWM fan headers. This is because PWM fans in conjunction with the motherboard chipset obtain temperature data from Digital Temperature Sensors on the CPU itself. All PWM fans are four pin, and if plugged into a conventional three pin supply will operate at full speed just like a three pin fan.

Fans can also be plugged into the power supply's 5 volt line instead of the 12 volt line (or between the two for a potential difference of 7 volts, although this cripples the fan's speed sensing) to run them at a reduced speedMost fans will run at 5 volts once they are spinning, but may not start reliably at less than 7 V. Some simple fan controllers will only vary the fans' supply voltage between 8 V and 12 V to avoid this problem entirely. Some fan controllers start the fan at 12 V, then drop the voltage after a few seconds.

Soft mounting fans (e.g. with rubber or silicone fan isolators) can help reduce transfer of fan vibrations to other components.

Intel has recently developed a piezoelectric fan for use in desktop PCs, which is quieter than motor fans and consumes a fraction of the power. Watercooling is a method of heat-dissipation by transferring the heat through a conductive material which is in contact with a liquid, most often demineralised water and an additive to prevent bacterial growth and provide cosmetic effects. This heated water travels in a loop which usually contains a reservoir, radiator and pump. Recent advances in 12v DC pump technologies (for the first time specifically geared-for pc development) allow for new pumps to be both extremely powerful and extremely quiet. Loops can be made up of any combination of these items and some aren't required such as the radiator or reservoir if alternatives methods are used. The radiator often uses one or more fans to air cool the radiator fins and dissipate the majority of the heat at this point.

The most common loop order is reservoir to pump, radiator then the watercooling block and back to the reservoir. The radiator and fan efficiency has the greatest effect on the noise level and cooling efficiency but watercooling is currently the most effective and potentially quietest methods of cooling above ambient temperatures.

There is an inherent danger in the use of water around electrical equipment and leak testing the loop is always recommended before attaching any parts to the motherboard, after all loop connections have been made. The 12v DC pump can be run using batteries or a power supply making sure no power is going to any other part of the system. Because of these risks and the use of water under pressure watercooling is a greater technical challenge to set up due to the number of components and case modification usually required.

Secondary Storage

Hard Drives

Figure: Silicone grommets in a computer case for mounting a hard drive to reduce vibration.

Previously, hard drives used ball bearing motors, but these generated excessive noise when the rotational speed of the drive was increased to 5400 RPM or 7200 RPM. More recent desktop hard drives use fluid bearing motors. The first hard drive widely reputed to be quiet was the Seagate Barracuda ATA IV.

The smaller 2.5" form-factor hard drives generally vibrate less, are quieter, and use less power than the traditional 3.5" drives. On the other hand, they often have lower performance and less capacity, and cost more per gigabyte.

To minimise vibrations from a hard drive being transferred to, and amplified by, the case, hard drives can be mounted with soft rubber studs, suspended with elastics or placed on soft foam or Sorbothane. Hard disk enclosures can also help reduce drive noise. Care must be taken to ensure that the drive gets adequate cooling. Hard disk temperatures can often be monitored by SMART software.

Solid-state Storage

Solid-state drive (SSD) storage offers faster seek times, lower power consumption, no moving parts; making it theoretically more reliable and silent. There is scope for very high capacity drives; however, they are currently extremely expensive per GB and are fairly scarce.

Compact Flash Cards

A few people use Compact Flash (CF) cards for storage. Because they use a slightly modified Parallel ATA (PATA) interface, a simple adapter is all that is needed to connect CF cards to function as an PATA or PC Card hard disk. CF cards are also small, allowing SFF PCs to be made, produce no noise, use very little power (further reducing heat output in the AC/DC conversion in the PSU), and an insignificant amount of heat. However, they are very expensive per GB and are only available in small capacities.

There are also issues regarding the maximum number of writes to each sector; often specified as 100,000 write cycles. However, there are industrial grade cards which specify a higher number of erase cycles, and different file systems, or technologies such as Enhanced Write Filter can reduce the writes to the card. Also, CF cards will fail gradually, so it will be easy to notice before any significant amount of data are lost, unlike the possible immediate failure of HDDs. Due to their small capacities they are easy to back up entirely, and often have 10 year or even lifetime warranties. Windows minimisation projects such as Winimize (Windows 98, less than 20MB) and Linux projects such as Puppy Linux mean that running an OS in small capacity, cheap compact flash card is possible. Because they have many OS components removed, they offer a smaller attack surface for malware to target.

The sustained transfer rate of current CF cards is a maximum of around 25 MB/s, compared to an average of around 70 MB/s for modern hard drives. However, the speed of flash memory is increasing at a faster rate than that of hard drives, and they have minimal seek times compared to hard drives, which increases the speed of loading many small files, and makes the PC seem more responsive as most operations performed by the OS include small files. Due to the fast seek times CF cards also don't show the effects of file system fragmentation like hard drives do.

Newer CF cards support faster transfer protocols like DMA. It is possible to use a Compact Flash card for storing only information that isn't changed very often, such as music, videos and binary executables, while storing the small configuration files and other frequently modified data on a small hard drive

Where a motherboard supports booting from USB drives, they can be used in a similar fashion to CF cards to run the OS. With some Linux distributions, it is not much harder than using a CF card. As they both use flash memory, they have the same advantages and disadvantages, however speed is limited by the USB bus.

Gigabyte i-RAM

The i-RAM is a solid-state disk which has four DIMM slots to allow regular PC RAM to be used like a disk. It is much faster than a hard disk, does not have the write cycle limitations of flash memory, however it requires power continuously in order to maintain its contents (from standby power or a battery when the system is off), uses more power than many laptop hard drives, has maximum capacity of 4 GiB, and is expensive.

Problems and Solutions

All forms of affordable solid-state storage offer relatively small capacity. They can be used as main storage for tasks which do not use large amounts of data or large programs, such as web browsing or word processing. Larger files and programs can be stored on a secondary hard drive which is only accessed when needed. Keeping the OS, often accessed files, and smaller programs on a solid-state drive means that the hard drives can be powered down much of the time. Network-attached storage, or NAS, is another alternative, allowing loud hard drives to be stored remotely. Small USB drives or CF cards can be used to make the process of network booting easier also.

Optical Drives

Optical drives can be slowed down by software to quiet them, such as Nero DriveSpeed, or emulated by virtual drive programs such as Daemon Tools to eliminate their noise entirely. Notebook optical drives can be used, which tend to be quieter, however this may be because they tend to run slower (typically 24x CD speed, 8x DVD speed). Some DVD drives have a feature, commonly called Riplock, which reduces drive noise by slowing the drive during video playback. For playback operations only 1x (or real time) speed is required.

External Components

Laptop computers typically do not have power supply fans or video card fans, and they use smaller hard drives. They also use many lower power components. However, laptop CPU coolers are usually smaller, so may be noisier than their desktop counterparts. Limited space, limited access and proprietary components make silencing laptops more difficult. A few laptops do not use cooling fans, for instance the Dell Latitude X1, Panasonic Toughbook W5 and T5, Fujitsu Lifebook P7120. Also, some netbooks, such as the Dell Mini 9, 10 and 12 do not have fans. The Mini 9 used SSD rather than a hard disk. The OLPC XO-1 has no internal moving parts.

CRT monitors can produce coil noise, as can the external power supply for an LCD monitor or the voltage converter for the monitor's backlight. LCD monitors tend to produce the least noise (whine) from when at full brightness reducing brightness using the video card does not introduce whine, but may reduce colour accuracy. An LCD monitor with an external power supply tucked out of the way will produce less noticeable noise than one with the power supply built into the screen housing. Dot matrix and daisy wheel printers are often noisy, and soundproofed boxes or cabinets can be used to reduce the noise. Another solution is to locate the printer away from the immediate work area or in another room, especially if it can be controlled through a local area network.

Chapter 5

Centrifugal Compressor

Centrifugal compressors, sometimes termed radial compressors, are a sub-class of dynamic axisymmetric work-absorbing turbomachinery.

Figure: Centrifugal impeller with a highly polished surface likely to improve performance

The idealized compressive dynamic turbo-machine achieves a pressure rise by adding kinetic energy/velocity to a continuous flow of fluid through the rotor or impeller. This kinetic energy is then converted to an increase in potential energy/static pressure by slowing the flow through a diffuser.

Imagine a simple case where flow passes through a straight pipe to enter centrifugal compressor. The simple flow is straight, uniform and has no swirl.

Figure: Jet engine cutaway showing the centrifugal compressor and other parts

As the flow continues to pass into and through the centrifugal impeller, the impeller forces the flow to spin faster and faster. According to a form of Euler's fluid dynamics equation, known as "pump and turbine equation," the energy input to the fluid is proportional to the flow's local spinning velocity multiplied by the local impeller tangential velocity. In many cases the flow leaving centrifugal impeller is near or above 1000 ft./s or approximately 300 m/s. It is at this point, in the simple case according to Bernoulli's principle, where the flow passes into the stationary diffuser for the purpose of converting this velocity energy into pressure energy.

Historical Contributions, the Pioneers

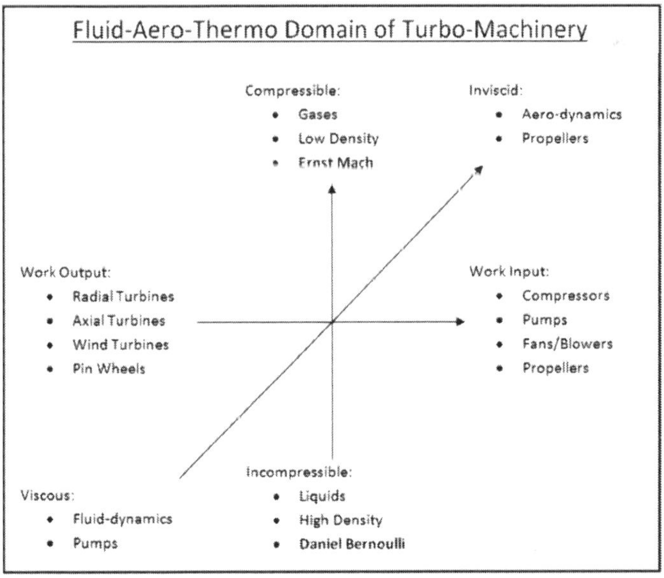

Figure : Aero-thermo domain of turbomachinery

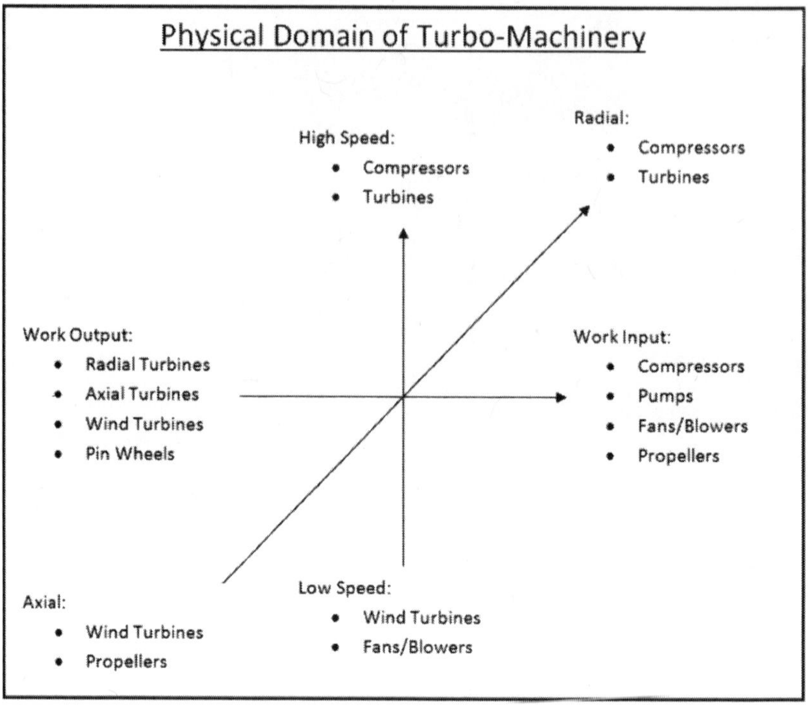

Figure : Physical domain of turbomachinery

Over this past 100 years, applied scientists like Stodola (1903, 1927–1945 Pfleiderer (1952), Hawthorne (1964), Shepard (1956), Lakshminarayana (1996 and Japikse (many texts including, 1997) have tried to educate young engineers in the fundamentals of turbomachinery.

These understandings apply to all dynamic, continuous-flow, axisymmetric pumps, fans, blowers, and compressors in axial, mixed-flow and radial/centrifugal configurations.

This relationship is why advances in turbines and axial compressors often find their way into other turbomachinery including centrifugal compressors. Improvements in centrifugal compressors have not been achieved through large discoveries. Rather, improvements have been achieved through understanding and applying incremental pieces of knowledge discovered by many individuals.

The horizontal axis represents the energy equation derivable from The First Law of Thermodynamics The vertical axis, which can be characterised by Mach Number, represents the range of fluid compressibility (or elasticity). The Zed axis, which can be characterised by Reynolds Number, represents the range of fluid viscosities (or stickiness).

Mathematicians and Physicists that established the foundations of this aero-thermo domain include: Sir Isaac Newton, Daniel Bernoulli, Leonard Euler, Claude-Louis Navier, Sir George Gabriel Stokes, Ernst Mach, Nikolay Yegorovich Zhukovsky, Martin Wilhelm Kutta, Ludwig Prandtl, Theodore von Karman, Paul Richard Heinrich Blasius, and Henri Coandã.

Again, the horizontal axis represents the energy equation with turbines generating power to the left and compressors absorbing power to the right. Within the physical domain the vertical axis differentiates between high speeds and low speeds depending upon the turbomachinery application. The Zed axis differentiates between axial-flow geometry and radial-flow geometry within the physical domain of turbomachinery. It is implied that mixed-flow turbomachinery lie between axial and radial. Key contributors of technical achievements that pushed the practical application of turbomachinery forward include: Denis Papin, Kernelien Le Demour, Daniel Gabriel Fahrenheit, John Smeaton, Dr. A. C. E. Rateau, John Barber, Alexander Sablukov, Sir Charles Algernon Parsons, Ægidius Elling, Sanford Alexander Moss, Willis Carrier, Adolf Busemann, Hermann Schlichting, and Frank Whittle.

Partial Timeline

<1689	Early turbomachines	Pumps, blowers, fans
1689	Denis Papin	Origin of the centrifugal compressor
1754	Leonhard Euler	Euler's "Pump & Turbine" equation
1791	John Barber	First gas turbine patent
1899	Dr. A. C. E. Rateau	First practical centrifugal compressor
1927	Aurel Boleslav Stodola	Formalized "slip factor"
1928	Adolf Busemann	Derived "slip factor"
1937	Frank Whittle	First gas turbine using centrifugal compressor
>1970	Modern turbomachines	3D-CFD, rocket turbo-pumps, heart assist pumps, turbocharged fuel cells

Turbomachinery Similarities

Centrifugal compressors are similar to axial compressors in that they are rotating airfoil based compressors as shown. It should not be surprising that the first part of the centrifugal impeller looks very similar to an axial compressor. This first part of the centrifugal impeller is also termed an *inducer*. Centrifugal compressors differ from axials as they use a greater change in radius from inlet to exit of the rotor/impeller.

Centrifugal compressors are also similar to centrifugal blowers of the style shown as they both increase pressure with the benefit of increasing radius. In contrast to centrifugal blowers, compressors operate at higher speeds to generate greater pressure rises. In many cases the engineering methods used to design a centrifugal blower are the same as those to design a centrifugal compressor. As a result they can at times look very similar. This relationship is less true in comparison to a squirrel-cage fan as shown. For purposes of generalisation and definition, it can be said that centrifugal compressors often have density increases greater than 5 percent. Also, they often experience relative fluid velocities above Mach number 0.3 when the working fluid is air or nitrogen. In contrast, fans or blowers are often considered to have density increases of less than 5 percent and peak relative fluid velocities below Mach 0.3.

Centrifugal compressors are also similar to centrifugal pumps of the style shown. The key difference between such compressors and pumps is that the compressor working fluid is a gas (compressible) and the pump working fluid is liquid (incompressible). Again, the engineering methods used to design a centrifugal pump are the same as those to design a centrifugal compressor. Yet, there is one important difference: the need to deal with cavitation in pumps.

Centrifugal compressors also look very similar to their turbomachinery counterpart the radial-inflow turbine as shown. While a compressor transfers energy into a flow to raise its pressure, a turbine operates in reverse, by extract in energy from a flow, thus reducing its pressure In other words, power is input to compressors and output from turbines.

Components of a Simple Centrifugal Compressor

A simple centrifugal compressor has four components: inlet, impeller/rotor, diffuser, and collector. Figure shows each of the components of the flow path, with the flow (working gas) entering the centrifugal impeller axially from right to left. As a result of the impeller rotating clockwise when looking downstream into the compressor, the flow will pass through the volute's discharge cone moving away from the figure's viewer.

Figure : Cut-away view of a turbo-charger showing the centrifugal compressor (blue) on the right end of the rotor

Inlet

The inlet to a centrifugal compressor is typically a simple pipe. It may include features such as a valve, stationary vanes/airfoils (used to help swirl the flow) and both pressure and temperature instrumentation. All of these additional devices have important uses in the control of the centrifugal compressor.

Centrifugal Impeller

The key component that makes a compressor centrifugal is the centrifugal impeller. It is the impeller's rotating set of vanes (or blades) that gradually raises the energy of the working gas. This is identical to an axial compressor with the exception that the gases can reach higher velocities and energy levels through the impeller's increasing radius. In many modern high-efficiency centrifugal compressors the gas exiting the impeller is travelling near the speed of sound.

Impellers are designed in many configurations including "open" (visible blades), "covered or shrouded", "with splitters" (every other inducer removed) and "w/o splitters" (all full blades). Most modern high efficiency impellers use "backsweep" in the blade shape. Euler's pump and turbine equation plays an important role in understanding impeller performance.

Diffuser

The next key component to the simple centrifugal compressor is the diffuser. Downstream of the impeller in the flow path, it is the diffuser's responsibility to convert the kinetic energy (high velocity) of the gas into pressure by gradually slowing (diffusing) the gas velocity. Diffusers can be vaneless, vaned or an alternating combination. High efficiency vaned diffusers are also designed over a wide range of solidities from less than 1 to over 4. Hybrid versions of vaned diffusers include: wedge, channel, and pipe diffusers. There are turbocharger applications that benefit by incorporating no diffuser..

Collector

The collector of a centrifugal compressor can take many shapes and formsWhen the diffuser discharges into a large empty chamber, the collector may be termed a *Plenum*. When the diffuser discharges into a device that looks somewhat like a snail shell, bull's horn or a French horn, the collector is likely to be termed a *volute* or *scroll*. As the name implies, a collector's purpose is to gather the flow from the diffuser discharge annulus and deliver this flow to a downstream pipe. Either the collector or the pipe may also contain valves and instrumentation to control the compressor. For example, a turbocharger blow-off valve.

Applications

Below, is a partial list of centrifugal compressor applications each with a brief description of some of the general characteristics possessed by those compressors. To start this list two of the most well-known centrifugal compressor applications are listed; gas turbines and turbochargers.

In their simple form, modern gas turbines operate on the Brayton cycle. Either or both axial and centrifugal compressors are used to provide compression. The types of gas turbines that most often include centrifugal compressors include turboshaft, turboprop, auxiliary power units, and micro-turbines. The industry standards applied to all of the centrifugal compressors used in aircraft applications are set by the FAA and the military to maximise both safety and durability under severe conditions.

Centrifugal compressors used in conjunction with reciprocating internal combustion engines are known as turbochargers if driven by the engine's exhaust gas and turbo-superchargers if mechanically driven by the engine. Standards set by the industry for turbochargers may have been established by SAE. Ideal gas properties often work well for the design, test and analysis of turbocharger centrifugal compressor performance.

Centrifugal compressors for such uses may be one- or multistage and driven by large gas turbines. Standards set by the industry (ANSI/API, ASME) result in large thick casings to maximise safety. The impellers are often if not always of the covered style which makes them look much like pump impellers. This type of compressor is also often termed an *API-style*. The power needed to drive these compressors is most often in the thousands of horsepower (HP). Use of real gas properties is needed to properly design, test and analyse the performance of natural gas pipeline centrifugal compressors.

Centrifugal compressors for such uses are often one-shaft multistage and driven by large steam or gas turbines. Their casings are often termed *horizontally split* or *barrel*. Standards set by the industry (ANSI/API, ASME) for these compressors result in large thick casings to maximise safety. The impellers are often if not always of the covered style which makes them look much like pump impellers. This type of compressor is also often termed *API-style*. The power needed to drive these compressors is most often in the thousands of HP. Use of real gas properties is needed to properly design, test and analyse their performance.

Because of the wide variety of vapour compression cycles (thermodynamic cycle, thermodynamics) and the wide variety of workings gases (refrigerants), centrifugal compressors are used in a wide range of sizes and configurations. Use of real gas properties is needed to properly design, test and analyse the performance of these machines. Standards set by the industry for these compressors include ASHRAE, ASME & API.

Centrifugal compressors for such uses are often multistage and driven by electric motors. Inter-cooling is often needed between stages to control air temperature. Note that the road repair crew and the local automobile repair garage find screw compressors better adapt to their needs. Standards set by the industry for these compressors include ASME and government regulations that emphasize safety.

Ideal gas relationships are often used to properly design, test and analyse the performance of these machines. Carrier's equation is often used to deal with humidity.

Centrifugal compressors for such uses are often multistage using inter-cooling to control air temperature. Standards set by the industry for these compressors include ASME and government regulations that emphasize safety. Ideal gas relationships are often used to properly design, test and analyse the performance of these machines when the working gas is air or nitrogen. Other gases require real gas properties.

Centrifugal compressors for such uses are often one-shaft multistage and driven by gas turbines. With discharge pressures approaching 700 bar, casing are of the barrel style. Standards set by the industry (API, ASME) for these compressors result in large thick casings to maximise safety. The impellers are often if not always of the covered style which makes them look much like pump impellers. This type of compressor is also often termed *API-style*. Use of real gas properties is needed to properly design, test and analyse their performance.

Performance

While illustrating a gas turbine's Brayton cycle, plots of pressure-specific volume and temperature-entropy. These types of plots are fundamental to understanding centrifugal compressor performance at one operating point. Studying these two plots further we see that the pressure rises between the compressor inlet (station 1) and compressor exit (station 2). At the same time, it is easy to see that the specific volume decreases or similarly the density increases. Studying the temperature-entropy plot we see the temperature increase with increasing entropy (loss).

If we assume dry air, and ideal gas equation of state and an isentropic process, we have enough information to define the pressure ratio and efficiency for this one point. Unfortunately, we are missing several other key pieces of information if we wish to apply the centrifugal compressor to another application.

A centrifugal compressor performance map (either test or estimated), shows flow, pressure ratio for each of 4 speed-lines (total of 23 data points). Also included are constant efficiency contours. Centrifugal compressor performance presented in this form provides enough information to match the hardware represented by the map to a simple set of end-user requirements.

Compared to estimating performance which is very cost effective (thus useful in design), testing, while costly, is still the most precise method Further, testing centrifugal compressor performance is very complex. Professional societies such as ASME (i.e. PTC–10, Fluid Meters Handbook, PTC-19.x) ASHRAE (ASHRAE Handbook) and API (ANSI/API 617-2002, 672-2007) have established standards for detailed experimental methods and analysis of test results. Despite this complexity, a few basic concepts in performance can be presented by examining an example test performance map.

Performance maps

Pressure ratio and flow are the main parameters needed to match performance map to a simple compressor application. In this case, it can be assumed that the inlet temperature is sea-level standard. Making this assumption in a real case would be a significant error.

- *Flow* — kg/s (range: 0.04 - 0.34 kg/s)
- *Pressure* — pressure ratio (t-t) (range 1.0 - 2.6 PR_t-t)

"t-t" implies the discharge total pressure is divided by the inlet total pressure (Pt_discharge/Pt_inlet).

As is standard practice, a horizontal axis labeled with a flow parameter. While flow measurements use a wide variety unit specifications, all fit one of 2 categories:

- *Mass flow per unit time*
- *Volume flow per unit time*

Mass flows, such as kg/s, are the easiest to use in practice as there is little room for confusion. Questions remaining would involve inlet or outlet (which might involve leakage from the compressor or moisture condensation). For atmospheric air, the mass flow may be wet or dry (including or excluding humidity). Often, the mass flow specification will be presented on an equivalent Mach number basis. It is standard in these cases that the equivalent temperature, equivalent pressure and gas is specified explicitly or implied at a standard condition.

In contrast, all volume flow specifications require the additional specification of density. Bernoulli's fluid dynamic principal is of great value in understanding this problem. Confusion arises through either inaccuracies or misuse of pressure, temperature and gas constants. Also as is standard practice, a vertical axis labeled with a pressure parameter. The variety of pressure measurement units is also vast. In this case, they all fit one of three categories:

- *The delta increase or rise from inlet to exit* (Manometer style)
- *The measured discharge pressure* (force)
- *The force ratio* (ratio, exit/inlet)

Constant Speed Lines

The two most common methods used for testing centrifugal compressors is to test along lines of constant shaft speed or along lines of constant throttle. If the shaft speed is held constant, test points are taken along a constant speed line by changing throttle positions. In contrast, if a throttle valve is held constant, test points are established by changing speed (common gas turbine practice). The most common method; lines of constant speed. In this case we see data points connected via straight lines at speeds of 50%, 71%, 87%, and 100% RPM. The first three speed lines have 6 points each while the highest speed line as five.

Constant Efficiency Islands

The next feature to be discussed is the oval shaped curves representing islands of constant efficiency. In this figure we see 11 contours ranging from 56% efficiency (decimal 0.56) to 76% efficiency (decimal 0.76). General standard practice is to interpret these efficiencies as isentropic rather than polytropic. The inclusion of efficiency islands effectively generates a 3-dimensional topology to this 2-dimensional map. With inlet density specified, it provides a further ability to calculate aerodynamic power. Lines of constant power could just as easily be substituted.

Design Point(s) *or* Guarantee Point(s)

Regarding gas turbine operation and performance, there may be a series of guaranteed points established for the gas turbine's centrifugal compressor. These requirements are of secondary importance to the overall gas turbine performance as a whole. For this reason it is only necessary to summarise that in the ideal case, the lowest specific fuel consumption would occur when the centrifugal compressors peak efficiency curve coincides with the gas turbine's required operation line. In contrast to gas turbines, most other applications (including industrial) need to meet a less stringent set of performance requirements. Historically, centrifugal compressors applied to industrial applications were needed to achieve performance at a specific flow and pressure. Modern industrial compressors are often needed to achieve specific performance goals across a range of flows and pressures; thus taking a significant step toward the sophistication seen in gas turbine applications.

If the compressor represent is used in a simple application, any point (pressure and flow) within the 76% efficiency would provide very acceptable performance. An "End User" would be very happy with the performance requirements of 2.0 pressure ratio at 0.21 kg/s.

Surge

Surge - is the point at which the compressor cannot add enough energy to overcome the system resistance or backpressure.

This causes a rapid flow reversal (i.e. surge). As a result, high vibration, temperature increases, and rapid changes in axial thrust can occur. These occurrences can damage the rotor seals, rotor bearings, the compressor driver and cycle operation. Most turbomachines are designed to easily withstand occasional surging. However, if the machine is forced to surge repeatedly for a long period of time, or if it is poorly designed, repeated surges can result in a catastrophic failure. Of particular interest, is that while turbomachines may be very durable, the cycles/processes that they are used within can be far less robust.

Surge Line

The Surge-line is the curve that passes through the lowest flow points of each of the four speed lines. As a test map, these points would be the lowest flow points possible to record a stable reading within the test facility/rig. In many industrial applications it may be necessary to increase the stall line due to the system backpresuure. For example, at 100% RPM stalling flow might increase from approximately 0.170 kg/s to 0.215 kg/s because of the positive slope of the pressure ratio curve. As stated earlier, the reason for this is that the high-speed line exhibits a stalling characteristic or positive slope within that range of flows. When placed in a different system those lower flows might not be achievable because of interaction with that system. System resistance or adverse pressure is proven mathematically to be the critical contributor to compressor surge.

Maximum Flow Line versus Choke

Choke - occurs under one of 2 conditions. Typically for high speed equipment, as flow increases the velocity of the flow can approach sonic speed somewhere within the compressor stage. This location may occur at the impeller inlet "throat" or at the vaned diffuser inlet "throat". In contrast, for lower speed equipment, as flows increase, losses increase such that the pressure ratio eventually drops to 1:1. In this case, the occurrence of choke is unlikely.

Choke

The speed lines of gas turbine centrifugal compressors typically exhibit choke. This is a situation where the pressure ratio of a speed line drops rapidly (vertically) with little or no change in flow. In most cases the reason for this is that close to Mach 1 velocities have been reached somewhere within the impeller and/or diffuser generating a rapid increase in losses. Higher pressure ratio turbocharger centrifugal compressors exhibit this same phenomenon. Real choke phenomena is a function of compressibility as measured by the local Mach number within an area restriction within the centrifugal pressure stage.

Maximum Flow Line

The maximum flow line, is the curve that passes through the highest flow points of each speed line. Upon inspection it may be noticed that each of these points has been taken near 56% efficiency. Selecting a low efficiency (<60%) is the most common practice used to terminate compressor performance maps at high flows. Another factor that is use to establish the maximum flow line is a pressure ratio near or equal to 1. The 50% speed line may be considered an example of this.

It is inappropriate to use the term choke in association with a maximum flow of all centrifugal compressor speed lines. In summary; most industrial and commercial centrifugal compressors are selected or designed to operate at or near their highest efficiencies and to avoid operation at low efficiencies. For this reason there is seldom a reason to illustrate centrifugal compressor performance below 60% efficiency.

Stonewall

Many industrial and commercial multistage compressor performance maps exhibits this same vertical characteristic for a different reason related to what is known as stage stacking.

Other Operating Limits

Minimum Operating Speed - the minimum speed for acceptable operation, below this value the compressor may be controlled to stop or go into an "Idle" condition.

Maximum Allowable Speed - the maximum operating speed for the compressor. Beyond this value stresses may rise above prescribed limits and rotor vibrations may increase rapidly. At speeds above this level the equipment will likely become very dangerous and be controlled to lower speeds.

Dimensional Analysis

The advantages between centrifugal compressors it is important to compare 8 parameters classic to turbomachinery. Specifically, pressure rise (p), flow (Q), angular speed (N), power (P), density (ρ), diameter (D), viscosity (mu) and elasticity (e). This creates a practical problem when trying to experimentally determine the effect of any one parameter. This is because it is nearly impossible to change one of these parameters independently.

The method of procedure known as Buckingham's Pi-Therom can help solve this problem by generating 5 dimensionless forms of these parameters. These Pi parameters provide the foundation for "similitude" and the "affinity-laws" in turbomachinery. They provide for the creation of additional relationships (being dimensionless) found valuable in the characterisation of performance.

For the examples below Head will be substituted for pressure and sonic velocity will be substituted for elasticity.

Π theorem

The three independent dimensions used in this procedure for turbomachinery are:

- M mass (force is an alternative)
- L length
- T time.

According to the theorem each of the eight main parameters are equated to its independent dimensions as follows

Classic Turbomachinery Similitude

Completing the task of following the formal procedure results in generating this classic set of five dimensionless parameters for turbomachinery. Full similitude is achieved when each of the 5 Pi-parameters are equivalent. This of course would mean the two turbomachines being compared are geometrically similar and running at the same operating point.

Flow coefficient:	$\Pi_1 =$	$\left(\frac{Q}{ND^3}\right)$
Head coefficient:	$\Pi_2 =$	$\left(\frac{gH}{N^2D^2}\right)$
Speed coefficient:	$\Pi_4 =$	$\left(\frac{ND}{a}\right)$
Power coefficient:	$\Pi_3 =$	$\left(\frac{P}{\rho N^3D^5}\right)$
Reynolds coefficient:	$\Pi_5 =$	$\left(\frac{\rho ND^2}{\mu}\right)$

Turbomachinery analysts gain tremendous insight into performance by comparisons of these 5 parameters with efficiencies and loss coefficients which are also dimensionless. In general application, the flow coefficient and head coefficient are considered of primary importance. Generally, for centrifugal compressors, the velocity coefficient is of secondary importance while the Reynolds coefficient is of tertiary importance. In contrast, as expected for pumps, the Reynolds number becomes of secondary importance and the velocity coefficient almost irrelevant.

Other Dimensionless Combinations

Demonstrated in the table below is another value of dimensional analysis. Any number of new dimensionless parameters can be calculated through exponents and multiplication. For example, a variation of the first parameter shown below is popularly used in aircraft engine system analysis. The third parameter is a simplified dimensional variation of the first and second. This third definition is applicable with strict limitations. The fourth parameter, specific speed, is very well known and useful in that it removes diameter. The fifth parameter, specific diameter, is a less often discussed dimensionless parameter found useful by Balje

1	Corrected mass flow coefficient:	$(\Pi_1 * \Pi_4) =$	$\left(\frac{m(Rt/k)^{0.5}}{pD^2}\right)$
2	Alternate#1 equivalent Mach form:	$\left(\frac{m_1}{\rho_1 a_1 D_1^2 \mathrm{Cos}(\alpha_1)}\right) =$	$\left(\frac{m_2}{\rho_2 a_2 D_2^2 \mathrm{Cos}(\alpha_2)}\right)$

3 Alternate#2 simplified dimensional form: $\left(\frac{m_1(t_1)^{0.5}}{p_1}\right) = \left(\frac{m_2(t_2)^{0.5}}{p_2}\right)$

4 Specific speed coefficient: $\frac{(\Pi_1)^{0.5}}{(\Pi_2)^{0.75}} = \left(\frac{NQ^{0.5}}{(gH)^{0.75}}\right)$

5 Specific diameter coefficient: $\frac{(\Pi_2)^{0.25}}{(\Pi_1)^{0.5}} = \left(\frac{ND^{0.5}}{(gH)^{0.75}}\right)$

It may be found interesting that the Specific speed coefficient may be used in place of Speed to define the y-axis, while at the same time the Specific diameter coefficient may be in place of Diameter to define the zed axis.

Affinity Laws

The following "affinity laws" are derived from the five Pi-parameters shown above. They provide a simple basis for scaling turbomachinery from one application to the next

$\left(\frac{Q_1}{Q_2}\right) = \left(\frac{N_1}{N_2}\right)$

From head coefficient: $\left(\frac{H_1}{H_2}\right) = \left(\frac{N_1}{N_2}\right)^2$

From power coefficient: $\left(\frac{P_1}{P_2}\right) = \left(\frac{N_1}{N_2}\right)^3$

From flow coefficient: $\left(\frac{Q_1}{Q_2}\right) = \left(\frac{D_1}{D_2}\right)^3$

From head coefficient: $\left(\frac{H_1}{H_2}\right) = \left(\frac{D_1}{D_2}\right)^2$

From power coefficient: $\left(\frac{P_1}{P_2}\right) = \left(\frac{D_1}{D_2}\right)^5$

Aero-thermodynamic Fundamentals

The following equations outline a fully three-dimensional mathematical problem that is very difficult to solve even with simplifying assumptions. Until recently, limitations in computational power, forced these equations to be simplified to an Inviscid two-dimensional problem with pseudo losses. Before the advent of computers, these equations were almost always simplified to a one-dimensional problem.

Solving this one-dimensional problem is still valuable today and is often termed *mean-line analysis*. Even with all of this simplification it still requires large textbooks to outline and large computer programs to solve practically.

Conservation of Mass

Also termed *continuity*, this fundamental equation written in general form is as follows:

$$\frac{\partial \rho}{\partial t} + \nabla\cdot(\rho v) = 0$$

Conservation of Momentum

Also termed the *Navier–Stokes equations*, this fundamental is derivable from Newton's second law when applied to fluid motion. Written in compressible form for a Newtonian fluid, this equation may be written as follows:

$$\rho\left(\frac{\partial v}{\partial t} + v\cdot\nabla v\right) = -\nabla p + \mu V^2 v + \left(\tfrac{1}{3}\mu + \mu^v\right)\nabla(\nabla\cdot v) + f$$

Conservation of Energy

The First Law of Thermodynamics is the statement of the conservation of energy. Under specific conditions, the operation of a Centrifugal compressor is considered a reversible process. For a reversible process, the total amount of heat added to a system can be expressed as $\delta Q = TdS$ where T is temperature and S is entropy. Therefore, for a reversible process:

$$dU = TdS - pdV.$$

Since U, S and V are thermodynamic functions of state, the above relation holds also for non-reversible changes. The above equation is known as the fundamental thermodynamic relation.

Equation of State

The classical ideal gas law may be written:

$$pV = nRT.$$

The ideal gas law may also be expressed as follows

$$p = \rho(\gamma - 1)U$$

where ρ is the density, $\gamma = C_p/C_v$ is the adiabatic index (ratio of specific heats), $U = C_vT$ is the internal energy per unit mass (the "specific internal energy"), C_v is the specific heat at constant volume, and C_p is the specific heat at constant pressure.

With regard to the equation of state. It is important to remember the while air and nitrogen properties (near standard atmospheric conditions) are easily and accurately estimated by this simple relationship, there are many centrifugal compressor applications where the ideal relationship is inadequate. For example, centrifugal compressors used for large air conditioning systems (water chillers) use a refrigerant as a working gas that cannot be modelled as an ideal gas. Another example are centrifugal compressors design and built for the petroleum industry. Most of the hydrocarbon gases such as methane and ethylene are best modelled as a real gas equation of state rather than ideal gases.

Pros and Cons

Pros

Centrifugal compressors are used throughout industry because they have fewer rubbing parts, are relatively energy efficient, and give higher airflow than a similarly sized reciprocating compressor or positive-displacement compressor.

Cons

Their main drawback is that they cannot achieve the high compression ratio of reciprocating compressors without multiple stages. There are few one-stage centrifugal compressors capable of pressure ratios over 10:1, due to stress considerations which severely limit the compressor's safety, durability and life expectancy.

Pros

Centrifugal compressors are often used in small gas turbine engines like APUs (auxiliary power units) and smaller aircraft gas turbines. A significant reason for this is that with current technology, the equivalent flow axial compressor will be less efficient due primarily to a combination of rotor and variable stator tip-clearance losses. Further, they offer the advantages of simplicity of manufacture and relatively low cost. This is due to requiring fewer stages to achieve the same pressure rise. Centrifugal compressors are less practical than axial turbines for propelling large aircraft, due to the resulting weight and stress, and to the frontal area presented by the diffuser.

Turbomachinery

Many types of dynamic continuous flow turbomachinery are treated in Wikipedia. Below is a partial list of these topics. What is notable about this set is that the fundamentals that apply to centrifugal compressors also apply to each of these topics. Certainly there are significant differences between these machines and between the types of analysis that are typically applied to specific cases. This does not negate that they are unified by the same underlying physics of fluid dynamics, gas dynamics, aerodynamics, hydrodynamics, and thermodynamics.

Compressors Research

Msu Turbo

MSU TURBO is a graphical user interface for a set of codes called GUMBO and TURBO used for analysis of unsteady flows in turbomachinery. MSU TURBO is used to define input parameters, and includes online documentation.

The codes are being developed at the Computational Simulation and Design Centre at the Engineering Research Centre at Mississippi State University (MSU). Development of the codes has been supported by NASA Glenn Research Centre, the US Army Research Office, GE Aircraft Engines, and Honeywell Engines Systems and Service.

GUMBO

GUMBO (Graphical Unstructured MultiBlock Omnitool), is a graphical preprocessor for defining grid and boundary condition information needed by TURBO. GUMBO is used to define grid block connectivity information, adjust the block sizes for parallel execution, and set boundary conditions at the block boundaries and interfaces.

Features of GUMBO

- Built-in help, online documentation and tutorial
- Data input and output
- Read/Write boundary condition information
- Read/Write grid block connectivity information
- Graphical boundary condition specification and diagnostics
- Graphical specification of boundary conditions from menu of boundary conditions

- Diagnostic detection of regions with unspecified or overlapping boundary conditions
- Automatic/Manual grid block connectivity
- Grid repartitioning for parallel execution
- Grid manipulation including smoothing, rescaling, translation, rotation, resizing and reordering.

TURBO

TURBO is a scalable, parallel multiblock flow solver for unsteady flows in turbomachinery. It solves the unsteady Reynolds-averaged Navier-Stokes equations with a decoupled k - epsilon

turbulence model. The code is being developed by Dr. Jenping Chen at Mississippi State University.

TURBO is parallelized with the use of MPI for data communication over multiple processors.

The parallel code supports arbitrary grid blocking which is useful in modelling complex geometries.

Features of TURBO

- Finite-volume discretization
- High-order fluxes using Roe's scheme
- Real & perfect gas models
- Implicit time marching using Newton & Gauss-Seidel relaxation
- Structured multiblock grids with arbitrary block connectivity
- Full annulus calculations (exact blade count)
- Partial annulus using phase lagged boundary conditions (single passage per blade row)
- Partial annulus using reduced blade counts
- Has been run on SGI, HP, CRAY, and Linux cluster computers.

Applications of TURBO

- Axial and centrifugal compressors and turbines
- Unsteady blade row interaction
- Wake-blade interaction (unsteady inlet/exit boundary conditions)
- Flutter simulations
- Rotating stall simulations.

Offshore Platforms

Tech Transfer has been involved in the design and analysis of reciprocating compressor installations on more than 150 offshore platforms. Over the last 15 years, we have completed the design and analysis for the largest reciprocating compressors installed on offshore platforms.

We have the ability to accurately perform the following structural deck and module analysis:

- stress analysis for lift
- transportation
- operation and storm conditions
- natural frequency analysis.

In addition, compressor operation can be simulated with unbalanced forces and structural vibration monitored throughout the platform.

Tech Transfer's compressor deck and module designs have led the offshore industry in the successful and reliable application of reciprocating compressors for Gas Lift, Sales Gas, Vapour Recovery, and Fuel Booster services.

FPSO Compressors

Tech Transfer has been involved in the design and analysis of reciprocating compressor installations on more than 20 FPSOs. For the last 15 years, we have completed the design and analysis for the largest reciprocating compressors installed on FPSOs.

We have the ability to accurately perform the following for any structural module or deck installation: stress analysis for lift, transportation, operation and storm conditions, ship's hull stiffness, natural frequency analysis, vibration with simulation of compressor operation, and fatigue.

Tech Transfer's compressor module designs have provided the FPSO industry with successful and reliable applications of reciprocating compressors for Gas Lift, Fuel Booster and Flash Gas services. These designs have been developed to solve past field problems on FPSOs, which are unique to this application. When Tech Transfer performs a reciprocating compressor module design and analysis, major pulsation and vibration problems are eliminated, which results in safe, reliable operation.

Tech Transfer is capable of full Reciprocating Package Design and Analysis services. Our engineers and design staff have extensive backgrounds with compressor manufacturers, compressor packagers and operators. We can complete your package design from equipment sizing and selection through fabrication drawings.

Tech Transfer's 20 year background in field vibration analysis has given us a vast amount of exposure to package design problems. Tech Transfer's expertise in package structural design and acoustical pulsation analysis has given us the ability to prevent pulsation and vibration problems during the design stages.

Tech Transfer has been an integral part of the package design for the largest separable, reciprocating compressors in the following installations: pipelines, fixed offshore platforms, floating offshore platforms and FPSOs.

Tech Transfer has been involved in the design and analysis of reciprocating compressor installations on many concrete foundations. For the last 15 years, we have completed the design and analysis for some of the largest reciprocating compressors installed on concrete foundations.

Tech Transfer is capable of designing foundations with pilings or drilled piers. We understand the pros-and-cons associated with skid mounting and block mounting compressors. To verify the design, foundation stress and dynamics are analysed with 3D FEA models that include concrete, rebar, anchor bolts, equipment, skids and soil stiffness.

Tech Transfer can supply foundation construction drawings and with bill of materials.

Tech Transfer is capable of full Reciprocating Package Design and Analysis services. Our engineers and design staff have extensive backgrounds with compressor manufacturers, compressor packagers and operators. We can complete your package design from equipment sizing and selection through fabrication drawings.

Tech Transfer's 20 year background in field vibration analysis has given us a vast amount of exposure to package design problems. Tech Transfer's expertise in package structural design and acoustical pulsation analysis has given us the ability to prevent pulsation and vibration problems during the design stages. Tech Transfer has been an integral part of the package design for the largest separable, reciprocating compressors in the following installations: pipelines, fixed offshore platforms, floating offshore platforms and FPSOs.

Tech Transfer has extensive experience in conducting acoustical pulsation analysis for reciprocating compressors and pumps. Our computer analysis results match field data. TTI's pulsation bottle and package designs will ensure acceptable field vibration with practical solutions.

Tech Transfer's success rate with pulsation analysis is unparalleled in the reciprocating compressor industry.

Tech Transfer initially developed its pulsation analysis technology to support its field vibration analysis of compressors. Today, Tech Transfer is a world leader in the pulsation and vibration analysis of reciprocating compressors. We perform pulsation analyses in complete compliance with API 618 and all of its options for stress and dynamic analysis.

Mechanical natural frequency analysis is performed with true Finite Element programs and the results match our field readings. Our knowledge of cylinder stiffness, bottle design and pipe support methods were developed from field vibration analysis. On many occasions, Tech Transfer has been given the opportunity to field-verify our pulsation levels and mechanical natural frequencies and our calculations match field readings.

Tech Transfer has met the most stringent demands of stress, dynamic and fatigue analysis required by major oil companies and certifying authorities. We respect such specifications and understand their purpose.

Structural skid and module designs for reciprocating compressors require both stress and dynamic analyses. Tech Transfer has the ability to complete stress analysis for lifting, wind loads, seismic conditions and sea motion. In order to prevent structural vibration problems from compressor unbalanced forces, Tech Transfer performs natural frequency analysis and simulation of operating vibration levels. Unbalanced forces are included from both mechanical and gas compression sources.

Tech Transfer has an extensive amount of experience with the design and analysis of platform decks and FPSO compressor modules. An accurate dynamic analysis of reciprocating compressor skids mounted on platform and ship decks cannot be completed without including these decks in the structural models. Tech Transfer has developed techniques for modelling and analysis that have been field-verified and match field readings.

ASME code recommends Section VIII, Division II, fatigue analysis of pressure vessels in services above 3,000 psig pressure. In addition, some certifying authorities require fatigue analysis of pressure vessels due to cyclic loading from sea motion on FPSOs. Tech Transfer can accurately complete such analyses and design practical modifications to meet these stringent requirements.

Reciprocating compressors are continually growing in size, horsepower and speed. If proper design and analysis are not completed, serious field problems can be encountered with concrete foundations. Tech Transfer can complete the foundation design, finite element stress analysis, finite element dynamic analysis, as well as construction drawings with specifications for any reciprocating compressor application.

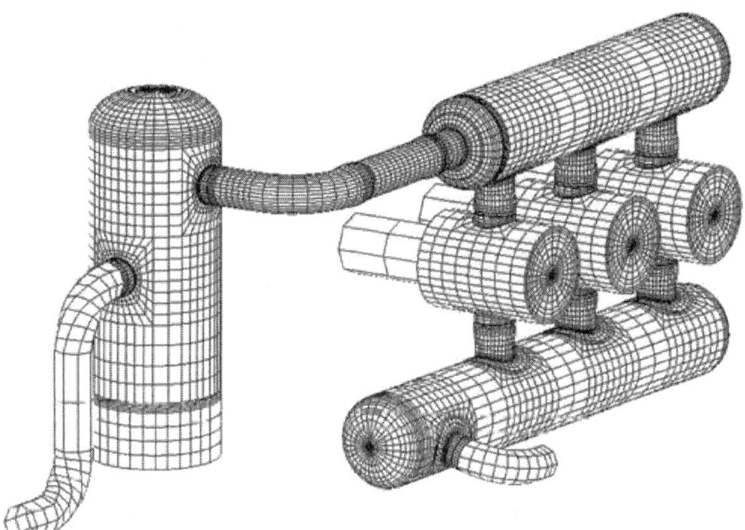

CAE Software for Axial Compressors

A meanline design and analysis tool for multistage axial turbomachinery including compressors, gas, steam, and hydraulic turbines, and pumps. *AXIAL* supports design-point and off-design analysis for subsonic and supersonic designs through an advanced pressure-based formulation. Other *AXIAL* features include:

- Estimate of hub and tip flow parameters

- Evaluation of compressor stall and choking conditions for both compressors and turbines
- Design analysis of multiple choked and stalled components, along with mass flow, pressure ratio, and power
- Setup of an arbitrary sequence of stage elements (rotor, stator, duct, IGV, EGV, etc.)

The Efficiency Workhorse

It's fairly safe to say that compressor technology, while mature in its basic function, is constantly changing. From new efficiency requirements to constant refrigerant phase-outs, this is one supplier category that will never be short of work. And let's not forget about noise: high performance, yes, but it better be quiet.

Fortunately, the industry's top suppliers have done their homework, and a little extra credit for that matter. Take the upcoming air-conditioning efficiency switch to 13 SEER as an example. While compressor makers recognise the need to do their part on the efficiency side, some have had the foresight to design for the additional effects the new regulations will have on their OEM customers.

"After the 13 SEER regulations take effect, it will be harder to differentiate equipment on efficiency alone," notes Frank Landwehr, vice president of marketing for Copeland Air Conditioning, a division of Emerson Climate Technologies (Sydney, OH, U.S.). "Other differentiating features will be essential to the OEM's ability to offer more value to the homeowner. Recent innovations in compression technology are helping customers offer this additional value."

Figure: Rheem partners with compressor manufacturers like Copeland and Bristol for high-quality, efficient, and dependable compressors for its full-line of condensing units and air-conditioners. Pictured is Rheem's Prestige Series® air-conditioner, which uses a Copeland Scroll compressor.

Emerson Climate Technologies says its Copeland Scroll UltraTech™ compressors, for example, are designed to increase system efficiency, but also reduce humidity, making it a great option for manufacturers that want to offer humidity control as a value-added feature.

These and other technology updates are giving air-conditioning and refrigeration OEMs all the right tools to make their equipment as efficient, quiet, and reliable as possible.

Energy Updates

It's no secret that the energy eater in any air-conditioning system is the compressor. So when the U.S. government required units to be 13 SEER by 2006, OEMs looked to their compressor suppliers for help.

According to Gary Bedard, vice president of Residential Product Marketing Management at Lennox International (Dallas, TX, U.S.), there are a couple of ways OEMs are attacking efficiency in the HVAC market. "You can make the heat exchanger surface more efficient, which basically adds more copper and aluminium. Or, the items that use electricity can be made more efficient themselves," he tells APPLIANCE. "For compressors, this can be done by increasing the efficiency of the mechanical parts of the compressor or by simply making the motor that turns the compressor more efficient. Compressor suppliers are really doing both to help manufacturers address the 30-percent increase in minimum efficiency levels by 2006."

For Emerson Climate Technologies, the answer was releasing a new version of its Copeland Scroll™ compressor designed specifically for the new guidelines. "With this significant industry event, scroll compression technology will be increasingly important to the U.S. market," Mr. Landwehr tells APPLIANCE.

The new design, which won an Innovation Award at this year's AHR Expo, is said to be 7-percent more efficient and 30-percent lighter than the prior version. To create the new compressor line, Emerson says its engineers made 20 design improvements, including an optimised scroll and seal design for improved efficiency, common shell/tube configurations and standardized parts for improved manufacturing and product quality, improved temperature protection, and a larger shaft bearing system for enhanced reliability. The unit is also 4-dBa quieter, according to Mr. Landwehr, thanks to a new quiet shut-down solution and a redesigned shell.

Rheem Air Conditioning (Fort Smith, AR, U.S.) attributes some efficiency gains to developments in two-stage compression and tighter manufacturing tolerances. "The two-stage compressor is another development that advances energy efficiencies because most often, it operates at a low speed (first-stage), and shifts to second-stage only during high-load conditions," explains Tim Hawkins, residential air-conditioning engineering manager. "In addition, refinements in compressor technology include tighter manufacturing tolerances for steel laminations and copper winding materials and other technologies that maximise efficiencies while reducing costs."

In commercial refrigeration, Danfoss (Baltimore, MD, U.S.) says variable-speed technology is

gaining popularity due to demand for greater refrigeration system efficiency and compliance with energy legislation pending in several U.S. states. As a result, Danfoss has introduced an a.c.-operated variable-speed hermetic compressor, type NLV, in 1,400, 2,000, and 2,600 BTUH versions.

The NLV units use a specially designed electronic control and optimised mechanics that allow a speed range of 2,000 to 4,000 rpm. According to Danfoss, the feedback loop from the temperature control, combined with self-learning logic in the electronics, allows the compressor to automatically match the appliance's load requirements with optimum speed, ensuring minimum energy consumption. This means that when cooling requirements change, like when a refrigerator door opens or contents are added to the cabinet, compressor speed is automatically adjusted to save energy. This is an improvement over single-speed compressors, which cycle frequently and can result in large variations in cabinet temperature and hence, energy inefficiency.

Reciprocating Advancements

Some industry professionals have proposed that reciprocating technology, which has commonly been used for its simple, low-cost design, will be outdated with the new 13 SEER minimum, leaving scroll as the leading technology. Bristol Compressors (Bristol, VA, U.S.), which makes both reciprocating and scroll compressors, doesn't believe this is always the case. "No one compression solution is optimal for all applications," Scott Hix, vice president of product engineering, tells APPLIANCE. "We believe that in applications of about 3.5 tons and below, the reciprocating [compressor] is the best solution and in larger tonnage systems, the scroll is the best choice."

According to Bristol, with the lower condensing temperatures and higher evaporating temperatures seen in the new 13 SEER systems, modernised reciprocating compressors are more efficient than scroll units at capacities less than 3.5 tons. The company notes that while scroll compressors have higher volumetric efficiency, reciprocating compressors have superior mechanical efficiency. This, it says, is because reciprocating products are less affected by friction, losing 23-percent less power to friction compared to a scroll unit.

Bristol also says volumetric efficiency is important because at lower condensing temperatures, the mass flow from reciprocating compressors will be higher than that of a scroll unit for equal system rated capacity. This, the company admits, results in some reduced system efficiency that must be made up in higher compressor efficiency, which Bristol claims has been accomplished with recent advances in reciprocating technology.

According to Mr. Hix, one example is Bristol's new H2IJ Benchmark reciprocating compressor, which features advancements in both efficiency and sound technologies. "The compressor operates in either one or two cylinder modes by simply reversing the direction of motor rotation," he explains. "With a displacement reduction of 60 percent (about 50-percent actual system capacity), this technology is able to run longer in the low-capacity mode, with fewer start/stop cycles. The result of this is that humidity can be maintained at lower levels, air stratification is reduced, and the indoor blower can be run at lower speeds, resulting in lower indoor sound levels."

To further reduce compressor sound, the shape of the H2IJ's shell was designed to minimise compressor sound radiation. The unit also utilises a patent-pending, internal suction muffler that reduces pressure pulsation inside the shell.

Refrigerant Developments

With the R22 phase-out coming up in 2010, HVAC companies are starting to design products with new refrigerants, mainly R410A. "The performance of R410A has proven to be more efficient than the current R22 refrigerant, so the bulk of redesign focus by York and our compressor manufacturers has been with R410A," says Cos Caronna, senior director of Engineering and Product Management at York UPG (Wichita, KS, U.S.). "However, we are not losing sight that there are still 4 more years of production with R22, and York plans to be competitive in both areas." Rheem Air-Conditioners is taking the same approach. In fact, the company says its contractor customers still prefer R22 refrigerants. "We'll continue manufacturing these products until the 2010 transition date," Mr. Hawkins tells APPLIANCE. "Our goal is to have the products that contractors need when they need them."

Mr. Bedard of Lennox believes that acceptance of R410A will grow as the industry transitions to the higher efficiency products, which can handle the new refrigerant better.

One future alternative that several compressor makers are looking into is CO_2 because of its environmental benefits. The refrigerant is already being introduced in regions like Europe, where there is discussion of banning HFC refrigerants all together. And while this green refrigerant may still be new to the North American market, it is a very real option for many global manufacturers.

Tecumseh Compressor Grup (Tecumseh, MI, U.S.) has been working with CO2 technology for several years and introduced its new EPACO2OL product at this year's AHR Expo. Although the CO2 compressor won't be commercially available until 2006, William J. Merritt, Jr., vice president of sales and marketing, says units are currently being field-tested around the world. In fact, Tecumseh has been supplying the U.S. military with CO2 compressors for heat pumps and air-conditioners throughout the last 3 years.

"CO_2 is something that has really been driven from the green standpoint," Mr. Merritt tells APPLIANCE. "I think a lot of the people in North America haven't addressed it because they are driven more by North American business opportunity first. We looked at this really from a global perspective, and the global demand is for more green refrigerants that can be used everywhere in the world."

Some companies in North America, however, are leading the charge toward the environment-friendly refrigerant. Mr. Merritt says corporations like Coca-Cola and McDonald's are partnering with their equipment suppliers on CO2 initiatives. Possible applications, he says, include 1/5- to 1-hp refrigeration products like beverage coolers and ice makers, as well as heat pumps and hot water heater applications.

Tecumseh plans to offer its EPACO2OL product in both reciprocating and rotary versions that will also be available as part of a complete refrigeration system. Mr. Merritt says that although CO2 compressors in themselves aren't quite as efficient as HFC units,

Tecumseh is making up for that in system components, creating an overall refrigeration system that is as efficient or more efficient than current models.

ACC Compressors has also been researching CO2 over the years and plans to offer a CO2 compressor by 2005. The Italian company has spent a lot of R&D time attacking the challenges of making CO2 an efficient, cost-effective option for OEMs.

According to ACC, challenges of working with CO2 include very high pressures (above 100 bar) and the need for a transcritical cycle. The company explains that the critical temperature of CO2 is as low as 31°C, which means that no condensation is possible above that temperature. In addition, the CO2 cycle is more sensitive to suction temperatures than HC and HFC cycles, and discharge temperatures can also be high. Because CO2 volumetric capacity is very high in comparison with HCs and HFCs, displacements are much smaller. According to ACC, the high pressure differences between compressor discharge and suction force the use of two-stage compression for low back pressure (LBP) appliances and imaginative mechanical solutions. ACC says it is successfully overcoming that challenge with some patent-pending solutions.

Even with several compressor designs in the works around the world, Mr. Merritt of Tecumseh admits that CO2 technology is still in its infancy and will most likely be a niche product. "It is not going to dominate the market," he says. "It will just be another option, but a green one that will work for all locations."

The Bigger Picture

While the majority of HVAC suppliers and OEMs in the HVAC/R industry have addressed the basics of quality, Mr. Bedard of Lennox says OEMs need to look at quality on a larger scale, including after the unit has been installed. This is the main reason Lennox decided to offer Copeland's Comfort Alert system as a product feature. The on-board diagnostics system, which was designed to work with Copeland Scroll compressors, helps installing contractors identify problems in the field.

"Misdiagnosis makes up a surprising number of the compressors that are changed during servicing," Mr. Bedard says. "No one is happy with this arrangement—it doesn't solve the consumer's problem and the contractor wasted valuable technician time in the summer months.

The Comfort Alert helps to correctly identify when the compressor is the cause of the problem and when it might be something else.

Mr. Landwehr adds that the Copeland diagnostics system is also one more way OEMs can differentiate themselves, making the contractors job easier and the customer's experience with a service call that much more enjoyable. Mr. Bedard seems to agree. "In the long run," he says, "this keeps a customer satisfied."

How to Read A Compressor Map - Turbo TecH

When it comes to building boost, compressor wheel selection is the make-or-break decision facing the pressure-craving enthusiast. Too often ego enters the equation and when starry-eyed enthusiasts run big wheels, performance is the loser.

First, you need to be logical and realistic. Second, you need a compressor map to get a ballpark idea of what wheel will best fit your application.

A compressor map is a snapshot of the wheel's efficiency laid out on an X-Y axis graph. Let's break down a map by the numbers and plug in some hypothetical applications. Going into this process you need to first know how much power you intend to make and have an idea of how much boost you expect to run. The mechanical limitations of your engine and its displacement would be deciding factors in these estimates.

Pressure Ratio

The pressure ratio along the X-axis represents the boost you intend to run. The numeric pressure ratio value is calculated by taking the expected psi, adding one atmosphere (14.7) and dividing that number by one atmosphere (14.7).

Airflow

The Y-axis represents airflow in pounds per minute. As a general rule, each pound of air generated represents 10 hp. We are talking flywheel horsepower, so if you, like myself, think in "wheel time," make allowances for driveline losses. It is along this line where you estimate engine power output.

Efficiency Island

Not fantasy island, so be real here. This is the bull's-eye of efficiency. The centre island is where the turbo is at its peak efficiency.

Wheel Efficiency

The outer rings of the graph represent reduced efficiency. The numbers and how much they drop off varies from compressor wheel to compressor wheel as does the corresponding loss of power potential.

Surge Limit

Any points plotted to the left of the Surge Line indicate the wheel in question is too big for the expected boost and power output. There would not be enough exhaust flow energy to spin the wheel fast enough to generate viable boost.

Overspin Choke

Any points plotted to the right of the graph mean the wheel in question is too small and will have to spin too quickly to make the expected boost/power. At these extreme wheel speeds, efficiency goes out the door because the wheel chops the air so badly that the pore density of the charge air will likely cause a dramatic loss of power.

Compressor Wheel Speed

This measurement, illustrated as lines across the graph, represents the shaft speed of the wheel. Remember, the faster it spins, the hotter the charge air. This is why they invented intercoolers

The missing pieces of the puzzle here are the engine's redline speed, the A/R of the turbine side and the volumetric efficiency of the engine. How fast an engine runs will impact the shaft speed of the turbo and when optimal efficiency is realised. The turbine side of the turbo will determine the spool-up characteristics and responsiveness of the unit. Turbine performance and engine redline speed are closely related, and together both of these factors will have a dramatic effect on the efficiency of the compressor wheel.

Further, an engine that has better volumetric efficiency, usually by way of cylinder head modification, will be able to flow more air and consequently require a different wheel than the same model of engine with no valvetrain tweaks. The compressor map is just an estimate that quickly indicates if the wheel in question will work based on two factors.

A map is not the end-all turbo-sizing tool. Turbonetics has a Turbo Match Spread-sheet program that takes into consideration the above listed items as well as fuel enrichment, intake temperatures and intercooler effectiveness to help pinpoint the proper compressor wheel for a given application. Charting A T04E 57-trim Compressor Wheel using a 15-psi max boost level the pressure ratio tallies 2.02. Plotting 2.00, we see peak efficiency will be realised when the wheel is moving 26 to 33 lb/min of airflow, which is capable of supporting 260 to 330 horsepower. If the engine in question is tuned to pump out 400 horses this T04E 57-trim wheel will need to see a rise in max boost to about 23 psi. Can the engine handle that kind of pressure? Does it have forged pistons and upgraded rods?

Even at this setting the efficiency is in the second ring. Also, look what happened to the wheel speed. It went from around 87,000 rpm to nearly 106,000. This will heat the charge air even more, working the intercooling system. While this wheel would get it done at 400 hp, a move to a bigger wheel may be better while also leaving some wiggle room for turning up the boost at a later date when the engine and/or the fuel system etc. are modified to handle the extra pressure.

Shaded Area

Peak efficiency for this compressor will be realised on an engine running 15 psi that is expected to generate 260 to 330 hp.

Dot-1 : Here we see where 15 psi and 300 hp plots; right in the peak efficiency island.

Dot-2 : With expectations of 23 psi and 400 hp, efficiency is bumped into the second ring. Wheel speed goes from 87,000 rpm at Dot 1 to more than 106,000 rpm here. Intercooling will be of prime importance. Can the engine handle 23 psi?

Dot-3 : Expecting 15 psi and 450 hp, we see the combination plot outside the map in the overspin choke area. This compressor is way too small. A better flowing wheel or the move to a bigger housing/ wheel combination is in order.

Compressor Technology for Penta West Pipeline

With this construction, the capacity of the Penta West Gas Pipeline, which is a key supplier of natural gas for Austria and bordering states, will be doubled. The pipeline branches off from the 245 km West Austria Gasleitung (WAG) Pipeline at Oberkappel, Upper Austria, and runs along the border between Austria and Germany to Burghausen in Bavaria.

Under a contract of approximately $US25 million, GE Oil & Gas will supply three ICL units, which are designed to offer operating and environmental benefits. Each ICL unit consists of a centrifugal compressor, high-speed motor, transformer and variable-speed drive. The scope of the contract also covers spare parts, training and turnkey installation.

The units will be manufactured and tested at GE Oil & Gas facilities in Le Creusot, France, and will be delivered to the project site in August 2010, with commercial service expected to start in early 2011.

OMV Gas Project Manager Hermann Lang said "We decided in favour of ICL technology because GE Oil & Gas is breaking new ground in emission technology and noise reduction. After careful consideration of the sensitive location Neustift, ICL seemed to be the ideal technology to be used for the first time."

A Tested Solution

The company found its solution in the Benchmark Scroll compressor from Bristol Compressors. "We ran all sorts of back-to-back tests on Benchmark and competing compressors," says Jeff Tucker, senior project engineer at York UPG. "Benchmark outperformed the competition in every area—capacity, efficiency, sound, and price."

To test the unit, York designed a high ambient test to simulate desert conditions in which temperatures reached 120°F (49°C) outdoors and 100°F (38°C) indoors. "We shut off the compressor, then turned it back on again, expecting the compressor to come back on within 5 minutes," explains Mr. Tucker. "Benchmark started immediately every time."

During all of its design projects, York always selects two brands of compressors—a primary and secondary model. In this case, Benchmark was the primary choice and ended up being the final solution. "We spent 18 months designing a new chassis, and Bristol was extremely responsive throughout the entire testing process," Mr. Tucker says. "They bent over backwards to adapt, change, and update the scroll compressor to custom fit our needs."

Motoring Ahead with Compressor Technology

A new oil-free compressor is slashing energy consumption, maintenance costs and footprint area by using only a single motor plus compressor component that revolves at up to 60,000 rpm. Based on the latest British and other leading-edge innovation, this new development trims back the cost of compressed air and puts pneumatics firmly back among the ranks of modern technologies appropriate for automated systems.

According to Jeremy Sykes, managing director of Compair, the new machine is a big step forward over conventional oil-free compressors, which use a pair of interlocking dry screws driven through a gearbox.

The 'Quantima' machine boasts an induction motor with a central rotor and two turbo air ends, the whole thing supported by magnetic bearings, both axial and radial, with sensors that check positions 120,000 times a minute. The rotor has one compressor turbine head on each end. The first compresses to 3bar, after which the air is passed through an intercooler, before being compressed by another turbine on the opposite end of the same rotor to typically 7 or 8 bar after, which it then passes to an aftercooler.

The company developed the Q-drive motor – now patented by Compair – with a partner, whom Sykes declines to name.

"The clever part is cooling it," he says, which is accomplished by a combination of a water jacket and air jets. He claims there are "very low losses in the motor [driven by a standard water cooled inverter] and elsewhere".

Another important attribute, he insists, is the lack of vibratory noise - only noise carried by the air, which is 69dBA, regardless of model. The models range from the 150kW Q-26, to the 300kW Q-52, all of which sit in the same sized enclosure, 2.4m x 1.6m x 1.85m - which he says is about half the size of competing units.

Furthermore, and perhaps more importantly, Sykes told us, a lot of compressors use a vast amount of power off load, whereas the new machines, he claims, use only 2.5% of full load power off load, amounting to 7kW for a typical 300kW machine.

The bottom line is that the company believes customers replacing an existing compressor with one of the new machines can expect to achieve energy savings of up to 25%. And the capital cost is similar, he adds. There are reduced running and maintenance costs, and the new machines come with a 10-year warranty. For those concerned about environment credentials, energy savings of just 10% of 300kW would be equivalent to taking 75 family cars off the road or planting 16,000 trees.

To ensure that efficiency is maintained, each machine is equipped with a predictive maintenance system called 'Q-life', which constantly monitors the compressors' performance. Moreover, each compressor connects automatically to a remote monitoring centre, supported by a network of engineers who undertake all servicing during the 10-year warranty period.

Pointers

* Compressor uses a single moving component, in the form of a motor armature, with a centrifugal compressor at each end, rotating at up to 60,000 rpm
* The rotor is supported in magnetic bearings that are constantly monitored, as is everything else on the machine
* As it consumes little power under no load, energy savings are said to be up to 25% relative to conventional machines

Air Compressor

With more customized trucks running the streets with air-adjustable suspension setups than ever before, there is often widespread misinformation given about the need to adequately equip your air-suspended vehicle.

Recently, we were able to get one of the fastest movers in the compressor market to stop running long enough to jot down some of the most common questions he receives regarding compressors and their proper use. Sammy Chu, president of Viair Corporation, is kind enough to lay it all out for us in pretty simple terms. If you haven't guessed it yet, Viair makes its own line of 12-volt automotive compressors that outperform just about anything in the market and takes the poundings that we mini-truckers dish out on a regular basis due to our demanding suspension systems. Read on to find all of your answers to your questions by one of the most knowledgeable persons in the world of compressor technology.

How do we choose a compressor that is fast enough for my setup?

The bigger the tank, the more airflow is required from the compressor to charge the tank. As a rule of thumb, for each gallon of tank capacity, a minimum of 0.15 cfm at 100 psi is recommended. This means if you have a 5-gallon tank, you need a compressor that puts out about a minimum of 0.75 cfm at 100 psi (0.15 cfm times five equals 0.75 cfm) for the compressor to recharge the 5-gallon tank in a reasonable amount of time.

Figure: Model 280

What is CFM?

The notation cfm (cubic feet per minute) refers to a volume of air produced by the air compressor at a given pressure or psi (pounds per square inch).

What size air tank should we use? Is one tank enough to suit my needs?

In general, tanks that are smaller than 3 gallons are normally used in air-assisted types of suspension setups. For full-function air-suspension systems where air replaces the use of conventional springs, a 5-gallon or higher volume tank is recommended. More than one large tank may work better to suit your needs, though.

You said to look at a compressor's flow rate at 100 psi, but why not look at a rate at 0 psi?

Well, cfm at free flow (zero head pressure) is a good reference point, but it does not really tell you the whole story. When you decide to purchase a car or truck, do you ever ask the salesperson how fast the engine will rev while standing still? It is also helpful to get tank fill rates from your compressor supplier.

Figure: Model 215

What is Duty Cycle?

A compressor's duty cycle determines the percentage of run time in a full cycle. A full cycle is the compressor run time plus rest time. Duty cycle can be a bit tricky in mobile pneumatics (automotive applications), since we are dealing with a wide range of conditions, applications, and ambient temperatures. Taking into consideration the above, manufacturers generally rate compressor duty cycles at 100 psi at a standard ambient temperature of 72 degrees Fahrenheit. This means a 20-percent duty cycle-rated compressor can run for five minutes (at 100 psi and 72 degrees) and then need to rest for 20 minutes. Five minutes of run time divided by five minutes of run time plus the 20 minutes of rest time equals a 20-percent duty cycle.

Figure: Model 300

How do we determine how much duty cycle we will need?

That's a really good question. This depends how often the compressor will need to come on and how long the compressor will need to run to fill your tank(s) from the cut-in pressure to the compressor cut-off pressure of your pressure switch. For example, let's say that a particular compressor takes two minutes to fill a 5-gallon tank from a cut-in pressure of 110 psi to a cut-off pressure of 145 psi. If we assume that the compressor is required to continuously run the fill cycle five times, then we are looking at 10 minutes of run time, so that is two minutes multiplied by five times. Let's say after you run the compressor continuously for 10 minutes, you would anticipate a rest time of about half an hour, then you are looking at a requirement for a minimum of 25-percent duty cycle-rated compressor. This is figured by taking 10 minutes of run time divided by 10 minutes of run time plus 30 minutes of rest time equalling a 25-percent duty cycle. Remember that duty cycle is rated at 100 psi, so the 25-percent duty cycle requirement that we have calculated based on a cut-in of 110 psi and a cut-off of 145 psi should be considered as the absolute minimum duty cycle requirement for that particular application. It is a good idea to go a bit overkill on your duty-cycle requirement, which means that you should buy a compressor that has a slightly higher duty-cycle rating than what you require.

There are quite a few 100-percent duty-cycle compressors out on the market today. Does a compressor that has a 100-percent duty cycle mean that I can run the compressor all the time, 24/7?

This is a often general misconception, and the answer is no, you can't. You have to remember that duty cycle is rated at 100 psi at an ambient temperature of 72 degrees Fahrenheit, but certainly a 100-percent duty cycle compressor can run for much longer periods of time than a 30-percent duty-cycle compressor under similar conditions.

Figure: Model 550

Does it matter where we install the compressor on the vehicle?

This depends on several factors, such as the available space, whether or not the compressor is a sealed or vented type, as well as personal preference and aesthetics.

Wait, you just mentioned sealed versus vented-type compressors. What's the difference? Can you elaborate more on those?

A compressor that is sealed is moisture- and dust-resistant. This type is obviously a compressor that has no vents. A vented compressor has vents throughout the motor housing and sometimes comes with a cooling fan in the rear of the motor.

So, a vented type of compressor runs a lot cooler then, right?

Exactly! A vented air compressor delivers a higher duty cycle than a sealed compressor, since heat buildup more readily dissipates through the vents.

If a vented air compressor will give me a higher duty cycle, then what's the advantage of even considering a sealed-type compressor?

The preference of a sealed-type compressor is because of the fact that in mobile pneumatics, compressors are likely to be exposed to outside elements.

High-performance 12-volt compressors are often sealed against moisture and dust to ensure a prolonged life for the compressor.

Figure: Model 400

OK, but when is a vented compressor more suitable for use?

Vented compressors can only be installed in an area that is protected from the elements, such as inside a truck's bed when the area is covered, like when you use a tonneau cover on your truck.

That makes a whole lot more sense. So, if we need to install our compressor, say, next to our framerail, we should only consider buying a compressor that is sealed against moisture and dust, right?

Yes, and it is a good idea to route an intake filter away from and up to an area where it will draw in cleaner and dryer air if you are able to have a remote intake filter from the manufacturer.

How can we tell whether the compressor is sealed or non-sealed?

Manufacturers normally specify if a compressor is sealed against moisture and dust, and in many cases, you can tell the difference by simply visually inspecting the compressor. If you see vents, cracks, or holes in the motor housing, then it's not a sealed compressor.

So, does that mean if we can't see any vents, the compressor is definitely sealed from moisture and it's dust-resistant?

No, not always. Sealing a compressor for moisture- and dust-resistance is a very complicated process. It involves sealing all joints that are not necessarily visible to the human eye.

Figure: Model 100

Can a compressor that is moisture- and dust-resistant be operated safely underwater?

When a compressor manufacturer specifies that a compressor is sealed for moisture- and dust-resistance, it does not mean that the compressor is water- or dust-proof. Operating a compressor while it is submerged in water or sand is one fast way to throw away your money, and it's always obvious to us as to why the compressor fails when we get it back for warranty purposes.

What does the term maximum-rated working pressure mean? Why should we be concerned with a compressor's working pressure?

Maximum working pressure is the top-end pressure a compressor can be safely operated to without damaging the compressor or shortening the life of the compressor. Always look at the maximum-rated working pressure of the compressor you have purchased when choosing a pressure switch to use in your suspension system.

Figure: Model 325

What about adjustable pressure switches? Are we able to use them on our suspension system?

We generally do not recommend the use of an adjustable pressure switch. This is simply because you may run the risk of setting the cut-out pressure at higher than the maximum-rated working pressure rated for your compressor, or for that matter, you may exceed the maximum-rated working pressure for your air supply tank. This could lead to a catastrophe that no one wants to imagine.

How do we find the maximum-rated working pressure of a compressor?

That's simple. Just ask your reputable compressor supplier

for the compressor manufacturer's compressor specifications. It will have that information on hand if it knows the product enough to be selling it.

Is the amp draw of a compressor something that we really need to be concerned with?

Yes, the lower the amp draw the compressor requires, the better the compressor is. As a rule of thumb, the maximum amp draw should be no more than 30 amps

Compressor Technology Evolves to Meet Today's Challenges

Compressors are the workhorses that pressurise gas, enabling it to be liquefied, transported and distributed via pipeline. Compressor technology is dynamic and evolving due to ongoing improvements in performance, efficiency, reliability and control, which are absolutely critical if gas reserves are to be developed and exploited on a practical and economic basis.

This is true not just for new units, but also with respect to upgrades of existing units in the field. Applying better materials, component designs, fabrication techniques and other advances to existing machines gives compressors powerful new capabilities, usually in the very same footprint and often with no need to alter supporting utilities and infrastructure.

While many different types of industrial compressors exist, the most frequently employed compressors for pipeline use are centrifugal design. Axial compressors are generally used for air service and in mixed refrigerant trains of LNG.

Centrifugal Compressors

Centrifugal compressors make use of the principle of centrifugal force. When gas enters the housing, a rotating impeller forces it outward to the impeller rim, thereby increasing the gas velocity and, in general, the energy content of the gas. From there a diffuser, with or without stationary vanes, converts the increased velocity to higher pressure. In general, gas pumping is performed by centrifugal compressors driven by gas turbines, steam turbines or electric motors. They can be capable in flow from 1,000 cubic metres per hour up to 400,000 cubic metres per hour. Axial-flow compressors

Axial-flow compressors are high-efficiency units used where there is a requirement for high flow rates but low pressure. They employ rows of airfoils to progressively accelerate and compress gas flowing through the unit. Multiple airfoil stages are placed along a rotor, matching up with corresponding stages of stationary airfoils in the compressor housing. Axial-flow compressors are used in applications such as catalytic cracking plants, air service, air separation, LNG, nitric acid and gas-to-liquid applications.

Compressor components

Compressors have a number of key components in common, regardless of their design:

- Shaft end-seals: To eliminate or minimise leakage of compressed gas or the entry of air into the compressor casing (in case of subathmospheric applications). Different types are used depending on the application.
- Bearings: Hydrodynamic tilting pad radial journal bearings are generally used, normally equipped with thermocouples to monitor bearing temperature. Thrust bearings are usually double-acting, tilting pad bearings with an equalising device, fitted with thermocouples for temperature monitoring and sometimes with load cells in high-pressure applications to measure axial thrust.
- Oil and gas seal systems: Lube oil systems provide lube oil to the radial and thrust bearings, to the gear box, and to the driver, except for some gas turbines. A seal oil system supplies filtered oil to the liquid film rings or to mechanical seals at required pressures and temperatures. Seal oil systems may be combined with the lube oil system.
- Control systems: Many manufacturers offer proprietary control solutions based on programmable logic control technology. For example, GE offers a uniquely integrated driver-compressor control system. It provides protection and monitoring to maximise the security of the compressor against potentially harmful surge conditions, while enhancing process efficiency

and availability. One system controls the entire compression train, auxiliaries and related process equipment.

Compressor upgrades

Modern materials, designs, fabrication techniques and testing procedures have greatly improved the performance and efficiency of compressor components. Upgrades provide an economic, easy-to-implement and timely way to accommodate different operating conditions such as changed process conditions, production increases, higher available power from the driver, or to optimise performance.

New Air Compressor Technology Launch

Compressed air manufacturer, CompAir, has rewritten the rules of compressed air generation by launching Quantima, a unique and dynamic oil-free compressor technology.

Having only one moving part spinning in a magnetic field, Quantimas revolutionary Q-drive compression assembly forms the heart of the new compressor design. Its unique rotor and direct-driven compression elements are levitated by active electromagnetic bearings to allow them to reach speeds of up to 60,000 rpm.

CompAir has a reputation for leading product innovation and this latest advance sets the benchmark for oil-free air compression, delivering a compelling range of customer benefits to help drive down production costs, reduce energy consumption and improve environmental performance. This results in a much quieter and more compact, variable-speed product that has no gearbox, no oil, and is engineered to deliver industry-leading reliability.

CompAir engineers have considered Quantimas energy performance at every stage of the design, delivering a compressor range that has the lowest off-load power consumption of any compressor on the market at just 2.5% of full load power, equating to only 7Kw for a typical 300kW unit.

This high energy efficiency means that customers replacing an existing compressor with a new Quantima unit can make energy savings of up to 25%, reducing overall running costs dramatically. In addition, a users environmental credentials are improved significantly with the potential to cut annual $C0_2$ emissions by the same amount as taking 75 family cars off the road or planting 16,000 trees*. Commenting on the launch of Quantima, CompAir Chief Executive,

Nick Sanders said "It is very rare that you can develop a product that will fundamentally change the way an industry thinks. With Quantima, this is what we have achieved. It has taken the vision of our entire business and the drive of our development team over the last 5 years, not to mention a significant investment and extensive field-testing, to tear up the rulebook and develop the world-beating, environmentally focused product that customers have been waiting for.

"In launching Quantima we have opened the door to the fast growing oil-free market and placed ourselves in the driving seat of this strategically important sector. Having brought this technological revolution to the global marketplace, we are now leading the industry and are determined to stay one step ahead."

As part of the Quantima revolution, Quantima customers will also benefit from Q-life, a state-of-the-art predictive maintenance solution. By constantly monitoring the performance of the individual Quantima unit, 24 hours a day, 365 days a year, performance is optimised whilst predicting the compressors maintenance needs to maximise uptime.

Each Quantima compressor connects automatically to a remote monitoring centre, supported by a network of service engineers that carry out all of the servicing during the 10-year warranty period that forms part of the Q-life package.

With Quantima, its Q-drive and Q-life assurance programme, CompAir has added a major product range to its comprehensive portfolio of products, making it the only manufacturer to offer a complete range of rotary screw, piston and vane compressors worldwide, which includes high-pressure, oil-free and portable models.

*Carbon emissions figures published equate to a calculation based on a user achieving energy savings of 10% on a 300 kW machine.

Dresser-Rand Invests "Supersonic" Compressor Tech

Dresser-Rand's funding will help satisfy Ramgen's requirement with the Department of Energy's National Energy Technology Lab to obtain private matching funds for development. In total, Dresser-Rand expects to pay Ramgen up to an estimated $49 million to fund Ramgen's development and demonstration of this technology and to obtain and exercise an option to purchase the assets of Ramgen plus a royalty dependent upon future equipment sales. Ramgen is developing game-changing compressor technology that applies proven supersonic aircraft technology to ground-based air and gas compressors. The four principal advantages of supersonic compression technology when compared to existing compressor technologies are that it can achieve exceptionally high compression efficiency; it is capable of producing very high single stage compression ratios; its simplicity and size make it less expensive to manufacture; and it offers the opportunity for significant waste heat recovery.

Full development of the technology is expected to take place over the next few years.

"Dresser-Rand has developed many of the technology and product breakthroughs in our markets, and manufactures some of the most advanced products available in each of our product lines," said Vincent R. Volpe Jr., Dresser-Rand's President and CEO. "We are excited about Ramgen's innovative technology.

This investment is consistent with our strategy to be the leader in technology in our industry. This investment is also a testament to our business model of strong cash flow generation, which allows us to pursue these types of bolt-on acquisitions."

Douglas N. Jewett, CEO of Ramgen noted, "Dresser-Rand is recognised by many scientists and engineers as the leading compression technology company in the world. They are the ideal partner for Ramgen in completing the development of our supersonic shock wave compression. Dresser-Rand has the credibility and capability required for the massive roll out that will be necessary to make a difference with CO2 emissions and climate change."

Both companies acknowledged the critical role that has been played by the Department of Energy through the National Energy Technology Laboratory ("NETL") in the development of this break through compression technology.

Doug Jewett noted, "NETL scientists were essential in focusing us on CO2 compression." Carl Bauer, Director of NETL, commented that, "NETL has worked with and supported Ramgen based on our technical assessment that their technology can make a significant contribution toward addressing our nation's energy needs. We are pleased that Dresser-Rand is now applying their expertise and resources to accelerate completing the development of this key technology and moving it into widespread application." "Ramgen's compressor technology has application in high mole weight services such as refrigeration and wet gas. It is also applicable to process air services. But a specific emerging market opportunity that Dresser-Rand is pursuing with Ramgen as the highest priority is in the CO2 compression market as applied to Carbon Capture and Sequestration ("CC&S")," said Brad Dickson, Dresser-Rand's Vice President and Chief Marketing Officer.

"Compressing CO2 is one of the most expensive parts of CC&S. The Ramgen CO2 compressor is projected to reduce the capital costs of CO2 compression by as much as 65% and produce a minimum of 25% operating cost savings. Reducing the added cost of CC&S for electrical rates is essential for the worldwide effort to minimise climate change.

"Ramgen's compressor technology allows us to configure a two-stage CO2 compressor for a pressure ratio of 100:1, while conventional technology will typically require eight-stages of compression," said Pete Baldwin, President Ramgen Power Systems.

"The input power to Ramgen's two-stage compressor will be comparable to the eight-stage conventional approaches. The all-in capital cost is expected to be one-third of the conventional approaches."

LG's Eco-Friendly Strategy & Market Outlook

Earlier this year, LG announced its worldwide 'Life's good when it's green' initiative to reinforce its commitment to the environment and energy savings.

The initiative focuses on sustainability through 'Eco-design and Eco-products,' including the reduction of hazardous materials and responsible electronics recycling. LG is taking proactive steps to address global climate change, pledging to reduce greenhouse gas emissions throughout the entire lifespan of its products by 30 megatons and in manufacturing by 150 kilotons per year by 2020.

As environmental regulations increase globally, consumers are demanding products with greater efficiency and lower energy consumption. It is expected that developed markets including Europe, North America, and Korea will see increased demand for eco-friendly appliances.

In addition, as this movement expands globally beyond these developed markets, home appliances are expected to become more and more energy efficient. For example, refrigerators will require compressor technology to minimise power consumption during daily operation.

Linear History

LG Electronics is a leader in innovative linear technology, developing the first Korean-made compressor in 1973 and the first market-ready 'linear compressor' in 2001. The company released an enhanced second-generation linear compressor in 2003. LG's dedication to providing improved efficiency and cost competitiveness led to the launch of a third-generation linear compressor earlier this year.

LG currently owns more than 700 patents domestically and internationally, and plans to form a task force to develop more innovative technology in this area.

Since manufacturing its first linear compressor in 2001, LG has included its compressor technology in more than 2 million units. At present, LG is capable of producing more than 1.5 million linear compressors per year.

Characteristics and Benefits of 3rd-Generation Linear Compressor

While conventional reciprocating compressors turn the motor's rotational movement into the piston's linear movement to compress refrigerant, in LG's linear compressors, the motor is directly connected to the piston and moves along a linear track to compress refrigerant.

This allows the inclusion of structures that have substantial influence on efficiency, for example, a separate structure between the suction and discharge systems, a low friction structure, and a linear motor designed to minimise end-coil loss. This has led to efficiency improvements compared with conventional reciprocating compressors. In addition, compared to reciprocating compressor, linear compressor has only one mating surfaces (piston) so it is advantageous for reliability. Furthermore, by developing soft-start technology, LG has substantially decreased start-up noise and the compressor is able to start at a lower voltage than reciprocating compressors. This makes it an ideal product for places with unstable voltage.

Compared to the first and second-generation linear compressors, third-generation linear compressors have achieved substantial improvements. For example, the third-generation linear compressor has an optimally designed piston diameter and stroke performance, which significantly improves the capacity modulation characteristics.

In terms of overall performance, by improving the capacity modulation characteristics, the third-generation linear compressor has decreased the energy consumption of refrigerators. This has been achieved by improving efficiency at low cooling capacity. Energy consumption improvements are expected to continue as even higher efficiency is achieved at high cooling capacity compared with inverter compressors.

In terms of noise, the third-generation linear compressor is outfitted with a pressure suppression pulsating silencer, helping to reduce noise generated inside the compressor. This contributes to minimising noise by decreasing vibration in refrigerator pipes when the compressor is attached.

In terms of control, the compressor automatically detects refrigerator load. This allows the operation of a self-modulating control system that adjusts cooling capacity without a control signal from the refrigerator.

LG's technology can provide energy solutions for refrigerator manufacturers who use inverter compressors as well as conventional reciprocating compressors.

In addition, optimal mechanical design has innovated materials and processing methods and strengthened cost competitiveness.

The third-generation linear compressor has been incorporated into LG's side-by-side refrigerators, with plans to expand the applications to bottom mount refrigerators designed for the European market.

Third-generation linear compressors in the large-sized FC series cover a range of 220-350 W ASHRAE cooling capacities, while medium-sized linear compressors will be launched next year.

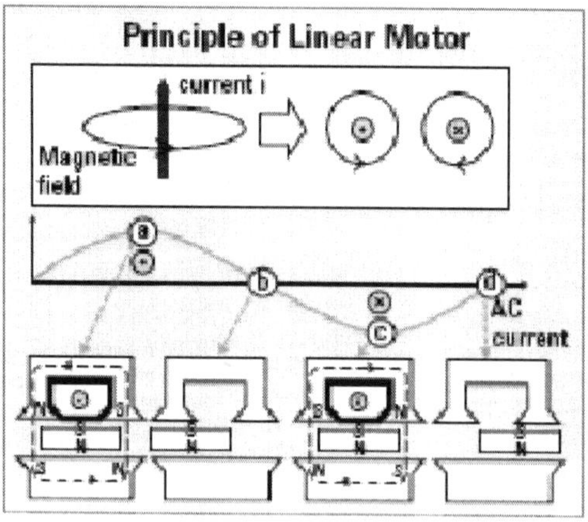

"LG's compressor strategy will allow development of the most energy-efficient compressors to quickly respond to stricter energy regulations," said Hyun Wook Lee, Director of the Compressor & Motor Business Team at LG Electronics. "We make every effort to improve refrigerator performance by developing top-level efficient compressors in a timely fashion, addressing rapidly strengthening energy regulations. With LG's unique linear technology, the company is focusing on designing next-generation linear compressors to set the global standard beyond the third-generation product."

Future Plans

In terms of LG's business domains, the third-generation linear compressor will allow the company to expand beyond its domestic market focus and establish strategic relationships with top-tier refrigerator manufacturers in North America, China, and Europe who demand high efficiency and large capacities. The company expects to expand its compressor business with the success of the linear compressor. LG is also committed to developing the next generation of linear compressors to respond to future energy battles.

With the launch of the third-generation linear compressor, LG expects to provide enhanced solutions in the competitive market of high-efficiency products.

Reciprocating Compressor Failure Due to Torsional Vibration

Torsional vibration problems in rotating machinery can be difficult to recognise. Unlike radial vibration, which can be easily measured with readily available sensors, torsional vibration is more difficult to measure because it involves the twisting of shafts while the machine is rotating. It can typically only be measured with special devices, such as strain gauges or torsional lasers. Due to the oscillating nature of reciprocating compressors, these machines can produce more torsional excitation than centrifugal machines. In addition, torsional problems might go unnoticed until something fails. In reciprocating compressors, this can be a coupling, crankshaft, motor shaft, etc. The purpose of this article is to present a case study of a reciprocating compressor train failure that was due to a torsional vibration problem. The failure, investigation and corrective action will be discussed.

Compressor Configuration

This case study pertains to a 4-throw, balanced opposed reciprocating compressor. The drive train consists of a 1,000 HP, 990 rpm induction motor directly coupled to a compressor through a flexible disc coupling. The drive train also includes a compressor-mounted flywheel assembly. The compressor cylinders are all double acting. This is a multi-service compressor. There are 3 stages of feedgas compression (1 cylinder per stage) and 1 stage (1 cylinder) of tail gas compression. There are no suction valve unloaders, and the only method for capacity control is recycle valves around both services.

History

The compressor was commissioned in 1994 and placed into continuous service. In 1999, there was a downturn in demand and the machine was operated at slightly less than 50 percent capacity.

Since the only means of capacity control was recycle, the power was the same as 100 percent capacity. Thus, the actual power was about 500 BHP higher than the power needed to meet the required duty. This was worth approximately $170,000 a year. Several ideas to reduce capacity and power were investigated. One idea was to replace the 1,000 HP, 990 rpm motor with a 500 HP, 500 rpm motor. However, if the business improved and more capacity was needed, the larger motor would then need to be re-installed. This was a relatively expensive option and did not allow a lot of flexibility for spot sales, since swapping motors would be a time consuming task. The other idea was either installing head end suction valve unloaders or simply removing the head end suction valves from all the cylinders. This would allow the compressor to operate with all the cylinders in a single acting crank end (SACE) mode and would reduce the capacity and power by approximately 50 percent. The business area and plant operations teams did not want to invest in automatic head end unloaders and preferred the option of simply removing the head end suction valves. Although this would require a compressor shutdown to install or remove the head end suction valves to change compressor capacity, it would not require a long shutdown. This work could be done in less than 8 hours. Rod loads and rod reversals in the SACE mode were determined to be acceptable. SACE operation also was discussed with the compressor supplier, and although the supplier did not typically consider this mode of operation, they did not have any concerns.

The head end suction valves were removed in May 1999. There were some minor changes in interstage feedgas pressures and temperatures, but all were well within design limits. The plant team also reported that the compressor linear vibration seemed higher, but the overall vibration levels were still at acceptable levels.

The machine ran without any significant issues until March 2002, when the tail gas connecting rod failed. This also damaged the crank shaft connecting rod journal, and the crank shaft and connecting rod were replaced. The machine then ran until October 2003, when the motor shaft failed at the coupling hub. The crack initiated from a keyway slot where there was a sharp corner that acted as a stress riser. The motor shaft was replaced and the compressor was put back into service. The original motor coupling hub was cast iron and was fastened to the motor shaft with a tapered lock mechanism. The shaft had a keyway that was not used (no key installed). The replacement coupling hub was steel and was fastened to the shaft with a key and an interference fit.

New Air Compressor Technology from Sullair

Australian air compressor technology group, Sullair Australia, has released an energy efficient VOC/VSD hybrid compressor. The VOC/VSD's variable speed drive (VSD) lowers power consumption at the top end of capacity and maintains constant system pressure whilst the variable output compressor (VOC) airend extends part-load energy savings.

Andrew Dove, Product Development Manager with Sullair Australia says the "driving force behind Sullair Australia wanting to combine the two technologies was a desire to achieve the greatest downturn, therefore providing customers with greater flexibility and efficiency." The VOC/VSD's unique fusion of VSD and VOC technologies delivers optimal airend efficiency for compressed-air applications with widely varying loads.

Individually the VOC has a turndown of 50 per cent; VSD as much as 30 per cent. The VOC/VSD hybrid air compressor can extend the range of downturn to 15 per cent while keeping the pressure stable and power consumption to a minimum. Through its ability to automatically vary both the motor speed and the compressor's airend length, the VOC/VSD's efficient operating range is broader than any comparable compressor.

The Sullair VOC/VSD utilises patented Optimal Flux Technology which allows precise and dynamic drive-to-motor calibration, delivering optimum performance and efficiency throughout the VOC/VSD's operating range. Optimal Flux Technology reduces motor temperatures by up to 11 degrees Celsius, thereby permitting greater turn-down, eliminating forced ventilation and improving motor efficiency and longevity.

The best way to determine if the VOC/VSD hybrid unit is the appropriate solution is by conducting an air audit. "If there are huge changes in demand at a particular site with for example, a peak shift in the morning and a light shift in the afternoon or night shift, the VOC/VSD may be beneficial in maximising your efficiency and energy savings" says Dove.

The VOC/VSD can be built to your specifications by Sullair's dedicated Customised Engineering team and fitted with a range of extras to suit every application. Designed with premium components and offering the lowest cost compressed-air available, the VOC/VSD has been engineered for long-term use and is supported by a range of extended service and

Centrifugal Compressor Thermodynamic Performance Monitoring and Analysis

Condition monitoring can be a very powerful tool to reduce maintenance costs while improving machinery availability and reducing spare inventory. Condition monitoring is different than traditional preventative maintenance techniques, which are time-based maintenance processes.

Condition-based monitoring involves regularly assessing the machine condition and taking action only if a potential problem is identified. One of the key parts of condition monitoring is thermodynamic performance monitoring and analysis. This involves collecting and reviewing operating pressures, temperatures, flow and power. Calculations also may need to be performed to determine machine efficiency and if there has been a loss of efficiency and/or capacity.

The purpose of this article is to describe one method for determining the overall efficiency of a centrifugal compressor. An example will be used to show how this technique was used to identify a performance problem.

Example

The example that will be used in this article is a 2,000 HP, 3-

stage integral gear type centrifugal compressor. It is used to compress atmospheric air in an air separation plant in Bountiful, Utah. There is an inlet filter upstream of the compressor and there are inlet guide vanes in the first stage to control capacity. There also is a flow meter downstream of compressor and the plant operator reported that the compressor power seemed high and the capacity low. The operator provided the following information:

- Inlet guide vanes = 100 percent open
- Machine flow = 425,000 SCFH (@ 70 F) (Measured on machine discharge)
- Discharge pressure = 93.2 psig (measured downstream of aftercooler)
- Pressure drop across inlet filter = 0.2 psig
- Ambient temperature = 80°F (also first stage suction temperature)
- Relative humidity = 30 percent
- Cooling water temperature = 71°F
- Second stage suction temperature = 92°F
- Third stage suction temperature = 88°F
- Measured motor amps = 204
- Measured motor voltage = 4,000
- Compressor first stage vibration = 0.4 mils p-p
- Compressor second stage vibration = 0.2 mils p-p
- Compressor third stage vibration = 0.5 mils p-p

Discussion

Compressors are seldom operated at the exact conditions the machines were designed for. As a result, it can be difficult to evaluate the overall compressor thermodynamic performance by directly comparing the actual operating conditions to the design operating conditions. One method to help with the evaluation is to determine the thermodynamic efficiency by calculating the thermodynamic power of an ideal process at the actual operating conditions and dividing this by the actual power. This calculation can be repeated at the original compressor design conditions, and comparing the actual efficiency with the design efficiency provides a better evaluation of the machine performance. However, there are many other factors, such as instrument tolerances and simplifying assumptions in the calculations, which will introduce some errors and cause some differences when comparing actual and shop tested efficiencies. Some of these issues will be discussed below.

In this example, the overall compressor performance will be evaluated by calculating the isothermal efficiency and comparing this with the calculated efficiency from the original shop performance test.

When a typical gas is compressed, the temperature increases. An isothermal compression process is one in which the gas is compressed and the temperature remains constant. So this would be the equivalent of having an infinite number of intercoolers, which represents an ideal case.

New Technology Helps Eliminate Compressor Moisture Damage

New IRIS® in-line scrubber technology from Multiphase Power and Processing Technologies (MppT) offers operators of reciprocating and turbo compressors in the pipeline and gas industry an effective, easy-to-install alternative to traditional methods of separating liquids from gas in pipelines.

If allowed to remain in the lines, moisture can cause erosion damage that can significantly affect performance in a matter of weeks. While a number of separation techniques have been used in the past, these methods are typically difficult to retrofit, or inefficient in their ability to protect equipment. The IRIS in-line rotary separator has an efficiency of 99.99 percent separation (down to five microns) and can be installed in a less than a day.

MppT is a joint venture formed by Aker Kvaerner and Dresser-Rand Company (D-R) in 1998. The new company combined the experience D-R has in designing and manufacturing reliable rotating machinery with Kvaerner Process Systems' expertise in packaging and separation technology. The objective was to develop and market proprietary expanders and other equipment to separate gases and liquids with the opportunity to generate power from previously wasted energy during oil and gas production.

The IRIS in-line rotary separator is a dynamic cyclone designed to work equally well in wellhead or pipeline applications to remove liquids from the gas flow in upstream pipelines. In these applications even small amounts of liquid can cause severe damage to compressors and reduce the service life of valves, liners, rings, rotors, and related equipment. The IRIS unit replaces, or may be used to complement, undersized existing vessel-based scrubbers.

In extensive testing conducted by MppT at the ChevronTexaco Ramirez Gas facility in Laredo, Texas, and at Hardman Resources at Woodada, West Australia, the IRIS unit achieved a separation efficiency of 99.99 percent over the design range. Testing was conducted at both the wellhead (in Laredo) and on the export gas line (at Woodada).

The IRIS separator is the most compact scrubbing device available, allowing it to be installed quickly with minimal alterations. The bi-phase device contains a rotary drum that uses gas pressure to spin liquids out of the gases. The IRIS assembly consists of three major components — an inlet housing, an exhaust housing, and a rotor/bearing system.

The inlet housing contains the swirl generating passages, liquid collection volute, inlet pipe flange, and inlet end bearing housing. The exhaust housing contains the diffuser/flow straightening vanes, the exit-end bearing housing, and the exit pipe flange. The rotor assembly consists of a drum pressed onto the shaft. Both inlet and outlet bearings are specially designed roller element bearings with separate, pressure-balanced seals to minimise process contaminants entering the bearings. A single flange keeps the inlet and exhaust housings together to form the pressure containment.

IRIS devices can be affixed to pipes from one to eight inches in diameter, making them suitable for most gas well applications. "The design is scalable, so producing other sizes is possible, but is dependent on the market," said Cody Novosad, IRIS product manager for MppT.

As an in-line device, retrofitting IRIS units to existing pipelines is easier than other methods of liquid and gas separation. "Because the IRIS separator is in-line technology similar to a valve or flow instrument, there's no additional piping or routing. In fact, just two flanges connect the device to the existing pipeline, so we've been able to install a unit in several hours without difficulty," Novosad explained. Because of the compact design and low weight, the unit does not require significant tooling or labour for change-out, and no additional maintenance is required for normal operation.

"The ease of a retrofit is even more significant when you consider that some operators have no liquid or gas separation system in place at all," Novosad said. "Their savings on maintenance would be remarkable and immediate."

In offshore gas compression applications where platform space is a premium, the innovative separator offers similar advantages. Traditional separation vessels are bulky and prone to problems such as foaming that can affect the system's sensors. "The IRIS in-line rotary separator is much smaller than traditional offshore separation systems, and it's more effective," Novosad said. "We have a prospective client operating in the North Sea who, because existing separator performance is inadequate, has spent millions of dollars on replacement equipment damaged by liquids. We've recommended the IRIS device as the obvious solution."

Although the IRIS separator was designed with the oil and gas market in mind, MppT recognises its potential for success in other markets, as well. "Because it's so effective at removing practically any liquid from a gas flow, it's just as applicable to air systems in the pharmaceutical, food processing, or any number of other process industries," Novosad said.

MppT has licensed the Dresser-Rand Product Services Division to be sales and service agents for the units throughout the Americas. "This agreement provides us with a great opportunity to increase sales potential and shows a commitment by Dresser-Rand in the products MppT has developed," said Greg Ross, general manager of MppT. MppT will use Dresser-Rand and Kvaerner Process Systems' offices worldwide to provide sales and service support for the IRIS technologies and the newer, larger RST packages.

IRIS® is a registered trademark of Multiphase Power and Processing Technologies (MppT).Dresser-Rand is a leader in energy conversion technology, with field-proven centrifugal and reciprocating compressors, steam turbines, expanders, gas turbine packages, and control systems. The company is positioned to deliver a complete package of solutions, from initial concept to equipment retirement for the oil and gas, chemical, and petrochemical industries

Chapter 6

Intage Compressors

Intage Compressors from Native Instruments was created in collaboration with Softube. The VC 76, VC 2A and VC 160 are apparently painstakingly accurate emulations of three of the most popular compressors in studio history. Each has its own sound, with the plug-ins adding the convenience of software side-chaining and parallel compression.

The VC 76 recreates famous UREI 1176 with a characteristic punchy and clear high end. Based on FET technology, the 1176 is renowned for its super-fast attack and unique 'All Button Mode' — a combination of attack, release, knee and ratio settings that creates a definitive, slightly overdriven sound. Using the 'no-button' mode you can apply the signature presence without compression.

The VC 2A is a software emulation of the ubiquitous LA-2A – an electro-optical tube compressor. Attack and release are determined automatically in response to the input signal. It's also easy to dial in the sweet spot, because the VC 2A has only two main controls: Gain and Peak Reduction, which controls threshold and the amount of compression. The VC 160 is based on a popular DBX VCA compressor found in studios all over the world. The VC 160 provides a continuously-adjustable threshold range, from 1:1 up to infinity. The VC 160 is mostly used for mixing drums, adding a definitive 'thwack' or 'knock' to kick and snare. It's also helpful as a mix-fix for troublesome bass, the VC 160 delivers additional attack and a gritty edge – suited to hip-hop and retro styles, but versatile enough to deliver results on a wide range of productions.

Air Compressor Test - Thrash Session

In life, more often than not experience will mean the difference between success and failure. Experience is a great tool to have when designing an efficient air suspension, too, but what do you do if you've never designed an air suspension before? Where do you turn for good information on the most important part of an air suspension, the air compressor? Magazine ads are great sales tools, but they usually don't provide all of practical information you'll need to make an informed decision, and brochures are only as good as the knowledge you possess to decipher their specs.

To clear the air (pun definitely intended!), we rounded up seven compressors and tested them to determine how fast each could fill a 2-gallon reserve air tank. We repeated this test several times to determine if the increase in operating temperature affected the compressor's output. We also pumped up the input voltage to determine the output of those compressors that could operate on more than 12 volts of power. And finally, we recorded the maximum amperage draw of each compressor to see if your mini-truck's electrical system is even up to the task of powering them. Our breakdown of each compressor will tell you how fast it will fill your air tank, how long it can operate, and how it will affect your truck's charging system.

Armed with this knowledge, you'll be able to pick the compressor that best suits your switch-hitting habits. Are you the type of person who only uses your air system to get you from Point A to Point B? Or are you hittin' switches at every stoplight and throwin' sparks on the freeway? Answer those two questions first and then you'll be able to select the right compressor for you. First, though, we must familiarize ourselves with all the of the important specifications that manufacturers use to rate their compressors. The compressor specs that matter the most are duty cycle, amperage draw, fill time, and airflow.

Duty cycle refers to the number of minutes an air compressor can operate at 100 psi, with an ambient temperature of 72 degrees Fahrenheit, divided by the number of minutes it must rest.

The equation looks like this:

$$X/(X+Y) = Z$$

X= compressor on time in minutes

Y= compressor off time in minutes

Z= duty cycle measured as a percentage

For example, a compressor that is rated as a 25 percent duty cycle compressor can operate continuously for 10 minutes at 100 psi and then must rest for 30 minutes. A 33 percent duty cycle compressor can operate for 15 minutes and must rest for 30 minutes. Everybody got that?

Current draw refers to the load (measured in ampheres) that the compressor places on the charging system of your truck. Amperage draw is an important spec because your truck's charging system needs to be up to the task of powering the compressor as well as all the other accessories. Technically speaking, the compressor, ignition system, audio system, and lighting system all run off of the battery, but it is the alternator that keeps the battery charged. If the output of the alternator cannot meet the demand for power that all of these accessories need, then the battery will die. Extended use will eventually burn out the alternator and destroy the battery and the compressors.

For example, if you install three compressors in your air system and each compressor draws 20 amps of current from the battery at maximum operating pressure, then a 60-amp load is being placed on the battery. Keep that number in mind because here are a few alternator output specs on four popular mini-trucks:

'98 Chevy S-10 2.2L four-cylinder engine, manual transmission

Alternator output: 100 amps

'00 Toyota Tacoma 2.4L four-cylinder engine, manual transmission

Alternator output: 80 amps

'95 Nissan Hardbody 2.4L four-cylinder engine, manual transmission

Alternator output: 60 amps

'86 Mazda B-2000 2.0L four-cylinder engine, manual transmission

Alternator output: 60 amps.

These are peak output figures. This means the alternator is not making this amount of power if your truck is parked and idling. The alternator will only make peak power output when you are driving it down the road and the alternator is spinning at more than 2,500 rpm. If you happen to own an '86 Mazda B-2000 or a '95 Nissan Hardbody, then the amperage draw of the three compressors is exceeding the output of your truck's alternator. This is bad news! Even when you are driving your truck down the road at 60 mph, the alternator cannot keep up with the current demands of the compressors, let alone the ignition system, lighting system, or that big audio/video system you have in the cab of your mini. Over time, this scenario will lead to a dead battery, a fried alternator, and possible damage to the compressors. When you choose a compressor, make sure you've factored in the other current-draining equipment installed in your truck such as neon lighting, HID headlights, audio amplifiers, and so on.

Careful planning will mean the difference between watching your headlights blink on and off whenever your air compressor turns on and having all of the electronic equipment in your truck operating at their full potential.

Airflow is measured in cubic feet per minute (cfm). Air compressor flow ratings are usually measured with 0 psi in the tank, which makes this specification very subjective. Measuring the airflow at 0 psi does you no good because the airflow at maximum operating pressure is what matters most. As pressure increases in the reserve tank, the airflow from the compressor decreases. So the true measure of airflow is the fill time. The quicker the compressor can fill your reserve tank, the more cfm of air it is pushing.

Fill time refers to the amount of time the compressor takes to fill up the reserve air tank in the system. The more efficient a compressor is, the faster it will refill the tank.

The Test

We took seven different compressors to our ultra top-secret testing facility, Shop X, and proceeded to put them through their paces. Our test equipment consisted of a switchable 12/24-volt power supply that is backed up with a pair of deep-cycle marine batteries. Our test bench features a built-in flow meter, a digital multimeter for testing current draw, a stopwatch, and a 2-gallon air tank. Each compressor was connected to the power supply using 4-gauge power wire and to the tank using 1/2-inch air line.

Our testing consisted of back-to-back runs during which each compressor filled the air tank to 150 psi. During the runs, we measured the operating temperature of the motor casing on each compressor using an infrared thermometer. We also measured the maximum current draw of each compressor using the digital multimeter and fill times using the stopwatch. Our target of 150 psi was established using a standard 145 psi pressure switch, which is widely used in most air suspension setups found on mini-trucks today. Here are the results and our evaluation of each compressor.

Viar rates the 325c compressor with a 13.8-volt input supply. The 325c draws a maximum 19 amps at a maximum working pressure of 150 psi. The 325c is also rated as a 33 percent duty cycle compressor when operating at 100 psi. Maximum airflow is 1.77 cfm at 0 psi and 0.76 cfm at 150 psi. The 325c weighs 6.95 pounds. A one-year limited warranty covers defects in workmanship and materials under normal operating conditions.

Compressor Viar 325C	*Current Draw*	*Fill Time*	*Temperature*
Test 1	16.7	2:57	87
Test 2	16.4	3:08	120
Test 3	16.3	3:13	138

Evaluation

The 325c worked very well with minimal current draw. This compressor would work very well in a multiple compressor install without pulling too many amps from your battery.

Viar rates the 400c compressor with a 13.8-volt input supply. The 400c draws a maximum 30 amps at a maximum working pressure of 150 psi. The 400c is also rated as a 33 percent duty cycle compressor when operating at 100 psi. Maximum airflow is 2.54 cfm at 0 psi and

0.71 cfm at 150 psi. The 400c weighs 8 pounds. A one-year limited warranty covers defects in workmanship and materials under normal operating conditions.

Compressor Viar 400C	*Current Draw*	*Fill Time*	*Temperature*
1.	26.4	1:30	78
2.	25.9	1:36	86
3.	24.4	1:44	95

Compressor Tech is the only machinery-oriented publication serving the natural gas industry. We cover the industry from the wellhead to the city gate, which means we are interested in wellhead and gathering compressors, booster stations, gas plants, storage facilities and mainline compressor stations. We also cover compression used in chemical and petrochemical process and refineries. The technology we cover includes the drivers, controls, auxiliary equipment systems and all related products and services necessary to design, build and operate a gas compression package. Focusing on the compressor helps ensure that the publication doesn't stray from its primary mission to the natural gas and process industries.

Simply put, that mission is to inform the operating market of new product designs and applications and to reflect back to vendors the significant current issues and trends in operation and maintenance from the field. Casual and comprehensible, we do that in a style that makes even the most technical subject accessible to all our readers.

The magazine has become even more important in recent years as the industry has undergone dramatic change. People change jobs more often today than in the past and we provide the kind of solid, useful information that helps them hit the ground running. We also provide an alternative to the kinds of traditional training activities that have suffered under the constraints of ever-tightening budgets and lean staffing. Dedicated readers rave that our unique content helps people them do their jobs better.

The publication has built a solid and tightly focused circulation of about 14,000. It is published 10 times a year and is distributed throughout the world. Being a controlled circulation publication, each reader must be involved directly with the design, application, fabrication, procurement, operation or maintenance of compressor systems, components or support systems to qualify for a free subscription. And we follow up on those readers on a regular basis to assure that they are still qualified to receive the magazine.

There's no doubt about it - CompressorTech^{Two} is the trade publication of choice for the vast majority of people working in the natural gas compression field. As readers, it's the magazine they depend on to stay informed. As advertisers, it's the magazine they depend on to reach the key customers who determine their success.

New Compressor Technology

On the table in a refrigeration and air conditioning wholesalers coffee room, the May/June 2006 Plumbing & HVAC Product News was lying open. On the page was an article titled 13-SEER. It led to a discussion with the mechanics sitting around the table. The group generated three questions: " I change compressors and do refrigerant charging; why should I worry about SEER? " Why should I care about SEER? Im in service side, not sales. " What is 13 SEER, heck whats a SEER?

SEERis an acronym for Seasonal Energy Efficiency Ratio. A single digit number, for example 13, indicates how efficiently an air conditioning system is working; the higher the SEER number, the greater the efficiency. SEER calculations are complicated and determined in a lab using very specific strictly controlled conditions. Simply, the calculation of SEER used in ducted A/C systems up to 65,000 BTU/Hour, includes: 1. Tightly controlled operating conditions for accurate comparisons. 2. Operating season length. 3. Operating cycles. 4. Operating under full and part load conditions. 5. Energy consumed during operation by the compressor and fan motors.

Why should refrigeration and air conditioning mechanics care about SEER? The simple answer & if it affects cooling and refrigerating systems they should care. The complex answer has factors that include: 1. Location and climatic conditions: Factors including altitude, prevailing winds, precipitation, humidity, enthalpy, temperature extremes and energy costs to name a few. In Canada, we have extremely different seasons, for example Ontarios weather is vastly different from Albertas. So a Toronto mechanic has different SEER concerns than his counterpart in Calgary. 2. Energy efficiency and cost effectiveness: the higher the SEER value the greater the energy savings, but at what cost? As R-22 continues toward the 2010 phase-out, we will concentrate on the realm of HFC refrigerants and for A/

C that means R-410A. This relatively new refrigerant operates at higher pressures and demands larger evaporators and condensers. This results in higher equipment costs for the end user. Cost effectiveness is a key consideration in both new and retrofit purchases. Why purchase a 13-SEER A/C system if the payout from energy savings never covers the added cost? (Government cash or tax benefits for purchasing energy efficient household appliances may change that.) All mechanics should be cognizant of these concerns and provide their customers with educated advice. 3. System balance: one truth that rings throughout all areas of HVACR training is system balance. The evaporator line, in red, is best matched with compressor 1, with the solid blue line. Remember that as the evaporators temperature difference Delta-T rises, its capacity increases. The compressor capacity increases with a rise in saturated suction temperature (SST). Thus, the evaporator and compressor 1 balance at 60,000 BTU/H, (five tons). The dashed blue-line of compressor 2 insects the evaporators line at a point that produces a lower capacity and will not meet the design conditions. The balancing act holds true for the condenser (plus heat of compression) the metering device, the refrigerant lines, and all other components through which the refrigerant passes.

Upgrading the compressor

Finally, if you are responsible for servicing systems that may need a compressor change, you have to worry about more than matching voltage, horse power and temperature range. Factors related to the new, high capacity, low energy, compressors will impact your replacement compressor choice. An R-410A air conditioning unit with an old ZP26K3E-PFV-230 Copeland scroll compressor. This compressor was designed for systems with 11+ SEER. Copelands new scroll, identified by K5E instead of the K3E in the older model, produces the same capacity with greater efficiency and lower noise. Other features include: 1. A small motor frame size. 2. An internal pressure relief valve (IPR). 3. Internal temperature protection. 4. Noise reduction on shutdown. Other compressor manufacturers are offering similar operating characteristics, so a good understanding of the new compressors and their added features is necessary for proper and effective compressor installation. Other concerns include: 1. Through capacity of shut off and reversing valves, filters and screens. 2. Suction line temperature gain. 3. The use of crankcase heaters, accumulators, and pump down cycles. 4. Compressor short cycling and minimum run times. 5. Never use a scroll compressor to produce a deep vacuum.

6. Oil type and charge. 7. Electrical components, controls and connections. Proper connections are a must. Also, make sure wires and other heat-affected components are not touching the compressor. 8. Electrical line variations and power interruptions should be avoided. 9. Reversed compressor rotation. Protection is inherent in most new scroll compressors, but the situation should be avoided.

Troubleshooting

Troubleshooting the new scroll/R-410A compressors follows the same rules and processes used on all compressors. Maintaining higher SEER equipment though, requires a little more time and attention to detail. All components must operate within the close tolerances set by the design criteria. All fans and ducts must operate at peak efficiency, meaning they are clean, leak free and properly installed. Refrigeration and air conditioning mechanics have to be aware of the new R-410A system characteristics. One refers to the need to respect the high R-410A pressures requiring special gauges and hoses. The other cautions the installer about excessive heat affecting components such as pressure relief valves and soft solder joints. More information about SEER, A/C unit capacities, testing, and design criteria can be found at the Websites of the various A/C and compressor manufacturers and HVACR associations and testing agencies.

Dresser-Rand Says New Centrifugal Compressor Represents Major Advance in Technology

In what it claims is a major advance in centrifugal compressor technology, Dresser-Rand said on Thursday that it successfully engineered, manufactured and tested the world's highest-density centrifugal compressor for re-injection service.

While the units will be used in the oil and has upstream market, the technology is applicable to other market segments requiring the compression of carbon dioxide and heavy gases such as urea production and carbon capture and sequestration.

The unit, which compresses a mixture of natural gas and carbon dioxide, is one of two trains that will be installed on a floating production, storage and offloading (FPSO) vessel destined for the "pre-salt" fields of offshore Brazil. Dresser-Rand booked the order in the first quarter of 2010, it said.

"We are extremely proud of this ground-breaking achievement," said CEO Vince Volpe. "The move reinforces Dresser-Rand's commitment to innovation and technological leadership. This advancement offers clients significant value, and complements our existing technologies in the global oil and gas industries. It represents a cost-effective solution that adds real value to our clients' capital projects and operations throughout the life of the equipment."

Chris Rossi, Dresser-Rand's vice president of technology and business development, said, "the design of this class of compressor presented many challenges such as reaching discharge pressures that exceed 550 bar (7975 psi) while compressing gases that are heavier than typical offshore natural gas blends, including a CO2-rich gas blend that is nearly 65% heavier than these natural gas blends. "At these conditions, the unit handles a gas in a super-critical state; or stated differently, a "gas" that is between its gaseous and liquid phases, which requires very accurate prediction tools."

The compressor was tested at full load and actual field conditions exceeding 550 bar discharge pressure and the range of specified gas compositions at the Company's facility in Olean, NY.

"The unit achieved a discharge pressure in excess of 560 bar (8120 psi) compressing the CO2-rich gas while exhibiting a very robust and stable rotor dynamic behaviour, generally considered the greatest challenge in this class of machinery. When the unit operated at this condition," added Rossi, "it reached what we believe to be the highest gas density any multistage centrifugal compressor has handled.

"This density level is equivalent to what a natural gas compressor would achieve at a discharge pressure of approximately 900 bar (13000 psi). Gas density is one of the most critical parameters affecting a compressor's stability and its ability to operate reliably," he said. Overall, the footprint required by a conventional compression and pumping module can be reduced by approximately 50% while its weight can be reduced by approximately 35% or more depending on the exact configuration, the company said. Officials noted that FPSOs, enhanced oil recovery and floating LNG operations were likely target markets for the compressors.

Automating Energy Efficiency in Compressed Air System Sub-Zones

Improving the Energy Intensity KPI

The objective of a compressed air management system is to improve the Key Performance Indicator (KPI) of "Energy Intensity" as it relates to compressed air. This translates into improving the ratio between the volume of product shipped and the net electricity (kWh) consumed by the compressed air system. How many thousand bottles, per compressed air kWh consumed, were shipped this month?

Compressed air management systems automate the continuous measurement and fulfillment of the appropriate volume, pressure, and quality of compressed air for each unique sub-zone in a compressed air system.

Energy Efficiency Awaits in the Sub-Zones of a Compressed Air System

The opportunities for energy savings often come from moving away from the traditional "one system for the whole plant" approach. Dusty Smith, P.E., (Director of Engineering, Pneu-Logic) commented, "Energy efficiency gains are unlocked when we find ways to optimally manage each sub-zone of a compressed air system." Compressed air management systems, like the Pneu-Logic PL4000, are designed to allow for the on-going measurement and optimal fulfillment of the appropriate volume, pressure, and quality of compressed air for each unique sub-zone in a compressed air system.

Compressed air systems assessments, focusing on the demand-side of the system, normally identify many different sub-zones with unique compressed air needs and requirements. Providing data that is normally not available to management, the system assessment will measure compressed air pressure, flow, and dew point, out on the factory floor where the machinery and processes are that use compressed air.

This is a learning process where engineers discover the different sub-zones in their compressed air system – the sub-zones where the energy efficiency opportunities are waiting!

A compressed air management system, like the Pneu-Logic PL4000, provides the ability to monitor and manage unlimited inputs from the compressed air system – permanently. The data inputs, from each sub-zone, turn the PL4000 into a powerful monitoring and trending system capable of providing management with the metrics they can manage. According to Eric Bessey, P.E., (Chief Project Engineer, Pneu-Logic), "Our technology differentiates itself with its ability to absorb data from a multitude of factory floor and compressor room inputs provided from sensors we install (typically flow, temperature, pressure, dewpoint, and kW). Our control algorithms are able to take this data and continuously direct the most energy efficient way to supply the demand for compressed air in all sub-zones of the system."

Compressed Air Pressure Sub-Zones

System assessments often identify processes requiring different compressed air pressures. Identifying these "sub-zones" and providing each with the appropriate pressure provide some of the biggest gains in energy efficiency. Dusty Smith, P.E., (Director of Engineering, Pneu-Logic) commented, "In plastic blow molding processes, the facility will use compressed air at both 600psig and 100 psig.

These two systems typically have a crossover point wherein 600 psig air can be expanded down to 100 psig air to augment low pressure air supply. Significant improvements in energy intensity metrics are realised by optimising the interaction between these two compressed air pressure sub-zones: one managed sub-zone for blow molding at 600-725 psi and another subzone at 100 psi for plant air functions."

"Another typical application is when systems are split between 30-40 psi air and 90-110 psi compressed air", continued Mr. Smith. "Food preparation plants are identifying energy-saving opportunities by supplying blow-off applications with lower pressure air supplied by blowers." Managing these two sub-groups of a system is made possible by the Pneu-Logic PL4000 compressed air system management technology.

Stabilising Air Pressure by Monitoring Signals from Storage Sub-Zones

Unstable compressed air pressure will reduce production output in a plant. It's like having an intermittent supply of electricity. To help stabilise plant pressure, compressed air storage and demand controllers are common recommendations coming from compressed air system assessments.

The PL4000 will take the inputs from pressure sensors at the storage tank(s) and flow and pressure inputs from the demand controller. The control algorithms can take this data from both sub-zones and select the optimum mix of air compressors to run.

If multiple compressor rooms are supplying air to the storage tanks, the Pneu-Logic PL4000 is able to centralise and coordinate the control of these air compressors located in different parts of the plant.

Eliminate Air Leaks by Shutting-Off Compressed Air Flow to Production Sub-Zones

Most industrial facilities have areas, supplied by compressed air, which do not run 24 hours a day, 7 days a week, all year-long. These areas however, due to compressed air leaks in the piping, normally consume compressed air 8,736 hours per year. Corey Farrens described the scenario, "A metal stamping plant we worked with has many different production lines. They wanted to shut-off compressed air supply to these lines when they weren't in production in order to save energy." The plant installed flow meters and electro-pneumatic gate valves (in each sub-zone) able to provide signals to the Pneu-Logic PL4000 on when to supply compressed air and when to open and shut the valves."

Managing and measuring compressed air flow heading out to different sub-zones is considered by many as, "the next frontier", in optimising compressed air systems. Many factories now understand their total compressed air costs because they are monitoring kW demand of their air compressors. The next step is to implement on-going measurement and management of compressed air flow heading to different sub-zones in the facility. In this way, different sub-zones can be accurately assigned kW costs that correspond to their production processes and thereby accurately manage their energy intensity.

Critical Use Sub-Zone: Compressed Air Used for Breathing Air

There are some applications, for compressed air, where lives literally depend upon safe supply of breathable compressed air. Mr. Smith described one such scenario "An enormous mine had stationary electric air compressors supplying compressed air to many, many zones. They experienced periodic power disruptions and wanted an automatic way to switch their breathing air supply over to the back-up diesel-powered air compressors." Along with managing the efficient supply of compressed air for the entire mine, the centralised PL4000 was able to provide this automated solution as an identified sub-zone.

Manufacturing facilities also use high quality compressed air in sandblasting and painting applications. Dew point and CO2 monitors are installed and can send alarm signals to the PL4000. The PL4000 will use this data to warn and ultimately shut-off the air supply and even the air compressors as a safety protection to personnel and manufactured product.

Air Quality Sub-Zones: Reducing Production Spoilage Rates

A fast way to reduce a plant's performance, with the energy intensity KPI, is to have product spoilage rates go up due to the presence of moisture and oil in the compressed air system. Direct-contact environments with food, delicate semiconductor chips, and surface painting processes have zero tolerance for specification deviations on pressure dew point, oil and particulate filtration. "Customers often ask us to measure, monitor, and manage air quality in critical process areas of their facility", said Mr. Bessey. "Dewpoint measurement devices are installed downstream and our management system provides visibility to engineering." "We also monitor differential pressure on the filtration products to ensure that the elements are changed in a timely manner to ensure that oil and particulate filtration specifications are met."

Some facilities operate separate oil-free and oil-flooded air compressed air systems. Many plants have both types of air compressors. Compressed air system management, with the PL4000, allows both air compressor systems to be controlled simultaneously.

Energy efficiency is also discovered when creating sub-zones for compressed air quality requirements. "It's common to see one dew point specification for a whole facility, based upon the highest air quality required in the plant – even if it's only 20% of the total air volume, " said Corey Farrens (Director of Sales, Pneu-Logic), "Our system assessments often recommend the installation of lower energy-cost refrigerated dryers for the portion of the factory that requires a +38 F dew point". The PL4000 compressed air system management product will monitor dew point to ensure that each sub-zone of the facility receives the quality of air it requires.

Cost-Effective, Open-Architecture, Non-Proprietary, and Configurable Technology

The PL4000 high-tech compressed air system management has matured over the past few years. Mr. Smith commented, "Communication flexibility is important. Our technology works extremely well with all protocols like Rockwell Automation's EthernetIP, and also with Modbus, CAN, and DeviceNet protocols Our engineers are also experienced in the details of how to write custom drivers for specialised IO scanners or to implement protocol bridges to get the myriad of air compressor controllers on the market, to communicate inwardly to the PL4000." The PL4000 SCADA server is ready for outward client connection and individual plants are given the "invisible codes", memory map and configuration files (as requested) to configure the system to the needs of their plant. There are also no technology licensing fees.

Innovation has also brought down the average selling price of the technology. Mr. Farrens said, "We can now customize the product to meet the needs of any customer. Solution prices can start at as little as $4,500. An innovative new payment option is also available. It allows customers to pay solely based upon realised energy savings. Pneu-Logic will provide a free system assessment and the installed PL4000 compressed air system management technology.

Dresser-Rand Introduces in-Line Electric Motor-Driven Compressor for Natural Gas Pipelines

For equipment operators in natural gas pipeline applications where quiet, emission-free operations are essential, Dresser-Rand has introduced the DATUM Model C electric motor-driven compressor.

The new pipeline compressor features an integrated, high-speed electric induction motor and the modular design advantages of the DATUM line. This unique design makes DATUM-C a solution for pipeline applications near urban areas where electric power is readily available, and environmental impact a concern.

The DATUM-C is the result of several years of research and development by Dresser-Rand in conjunction with a major U.S. pipeline client and the U.S. Department of Energy.

"The idea really came about through our discussions with one of our clients," said Harry Miller, product manager, Marketing at Dresser-Rand. Miller is one of the key engineers responsible for the original DATUM design.

"They sought to build a pipeline station in an underground bunker. This would require very little maintenance support and would be non-intrusive to the surrounding landscape. In addition, the client recognised there would be inherent security advantages with that type of installation. The challenge was designing a totally enclosed compression system with inline flanges and a natural gas-cooled electric motor."

As the early designs proved promising, D-R's project caught the attention of the U.S. Department of Energy. In 2003, D-R secured funding as part of a larger initiative for U.S. manufacturers to develop leading edge technologies for infrastructure reliability in the natural gas industry. "The DOE recognised the value of the DATUM-C, and the positive impact it could have on the pipeline industry. Their help has been instrumental in bringing this product to market."

DATUM-C offers the same advanced technologies and modular design concepts available in traditional DATUM compressors, providing equipment operators with greater ease of maintenance and life cycle value. "It's really the first spin-off technology based on the success of the DATUM compressor product line," Miller said.

In the early stages of developing the DATUM-C, D-R was presented with the challenge of selecting an electric motor suited for the project. After extensive research to find a motor company with a proven track record and technological capabilities, D-R selected Curtiss-Wright

"Once we identified Curtiss-Wright and saw some of the things it had accomplished for the U.S. Navy, we knew we had a good fit," Miller said. "There are other great industrial motor manufacturers out there, but Curtiss-Wright had the best combination of high tech motor and engineering expertise for this project."

The integrated two-pole induction motor from Curtiss-Wright uses magnetic bearings supplied by Kingsbury, and a self-contained natural gas cooling system. The unit operates at 10 MW with 6,900 volts available at the terminals, and a maximum continuous speed of 12,000 rpm.

While the natural gas pipeline industry was at one time largely limited to rural areas, in recent decades the ever-expanding network of pipes and stations has become commonplace in high-population centres. As a result, gas providers have had to balance the demands of a reliable supply of natural gas with the environmental and aesthetic concerns of their neighbours. These companies have continually sought new and innovative ways to reduce noise and emissions from their pipeline operations.

Originally, the proliferation of pipeline stations in remote areas required equipment that operated on the natural gas supply, and was not reliant on electric power. "Now with far more pipelines and stations, electric power is more readily available," Miller said. "And many industrial users can negotiate their own costs for electricity, so today the electric motor-driven compressor is even more practical."

The D-R DATUM-C compressor has a number of design features that make it uniquely suited to today's natural gas pipeline installations. The inline piping flanges of the DATUM-C allow the entire compressor to be installed directly into the line in both indoor or outdoor locations, greatly reducing infrastructure costs for new installations and for retrofit applications. The small footprint — approximately eight by 12 feet — and modular design also make installation and maintenance easier.

"The inline design makes the DATUM-C practical and efficient as a pipeline booster, and because of the compact size and quiet operation it would not even require the construction of a traditional pipeline station," Miller explained.

For pipeline applications, the DATUM-C is designed with a single-stage impeller for an inlet flow of 600 million standard cubic feet (mmscf) per day, and inlet pressure of 650 psig at 60 to 80 F. The discharge pressure is 1,160 psig. A second generation DATUM-C design is already in development with up to five impellers for higher pressure requirements.

"Most pipelines are operating at pressures of 1,000 to 1,200 psig," Miller noted. "But some companies are planning new pipelines at pressures up to 2,500 psig to transmit more gas in a given diameter. The DATUM-C will be able to meet those requirements, as well."

One of the most significant design features of the DATUM-C is the use of magnetic bearings. While magnetic bearings have been around for many years, previous systems relied on older analogue control systems, making the bearings difficult to tune and adjust, according to Miller.

"Previously, the magnetic bearing control system was designed, built and tuned on the test stand, but the only way to adjust it was to change the resistors and capacitors. Then if something changed in the field, one had to do it all over again, making it very difficult to manage, time consuming and expensive.

"But with today's digital controls, the computer-based algorithms make tuning very simple and quick. Now that operators have become more comfortable with the advanced, Windows-based digital controls, magnetic bearings will continue to gain even wider acceptance."

Because of the magnetic bearings, there is no need for an oil lubrication system, oil film seals or dry gas seals. "The unit is a completely closed system," Miller emphasized. "With no gas turbine or gear box, the entire system is extremely quiet and produces no on-site emissions. So it's very environmentally friendly."

The DATUM-C provides variable speed operation and can remain pressurised during shutdowns.

According to Miller, the DATUM-C compressor is already drawing favourable response from pipeline clients, and additional frame sizes are planned. "Because of the scalability features of DATUM, an entire family of units could be developed in the 5MW to 20MW size range to cover a variety of pipeline applications."

Dresser-Rand is among the largest suppliers of rotating equipment solutions to the worldwide oil, gas, petrochemical and process industries. The company operates manufacturing facilities in the United States, France, Germany, Norway, India, and Brazil, and maintains a network of 24 service and support centres covering 105 countries.

Air Compressor Designed for Tradesmen and Professional Finish Woodworking Crews

PORTER-CABLE unveiled a new 17 gallon air compressor, model C6001, with a vertical tank design, increased stored work capacity and an updated shroud for improved performance. The compressor, designed for tradesmen and professional finish woodworking crews, was introduced during the 2007 Association of Woodworking & Furnishing Suppliers® (AWFS) Fair from July 18 to July 21 at booth #5401 in the Las Vegas Convention Centre in Las Vegas.

Delta Porter-Cable

Recognising that floor space is at a premium, the 17 gallon vertical portable compressor tank, which includes a water drain valve, has a smaller footprint than horizontal tanks to take up less space and can be easily stored when not in use. The C6001 is factory-assembled with two wide-tread wheels for excellent stability and to allow the compressor to be easily moved.

For added convenience, the shroud design features a no-maintenance, oil-free pump and low amp, 120-volt universal motor that provides reliable starts in cold weather or with an extension cord. In addition, the C6001 is equipped with a console cover to protect controls and vital components; an easy-to-grip, protective handle; a large 2-inch air regulated pressure gauge for easy visibility; and a built-in cord wrap.

Ideal for the extended application of tools requiring low to medium airflow and for shorter burst tools requiring high burst airflow, the C6001 features 3.3 standardized cubic feet per minute at 90 pounds per square inch for fast recovery time and 150 max pounds per square inch for more stored work capacity and longer air tool performance. The C6001 will be available for purchase in the third quarter of 2007 for $219.

About Porter-cable and Delta Machinery

With a combined 189 years of woodworking experience, PORTER-CABLE and DELTA Machinery design and manufacture the industry's most complete line-up of portable, stationary and benchtop woodworking power tools, equipment and related accessories for professional woodworkers. Together, the brands have a history of industry-changing innovation, including introducing the world's first helical drive circular saw, first hand-held belt sander, first motorized miter saw, first portable belt sander, and first 10-inch tilting arbor saw.

Acquires Quincy Compressor

Atlas Copco has completed its acquisition of Quincy Compressor from EnPro Industries. The acquisition includes facilities in Bay Minette, AL; Quincy, IL; and in Kunshan, China, however, the acquisition of Quincy's Chinese operations is still pending, while the company awaits regulatory approvals from relevant local authorities.

The acquired business had 2009 revenues of $125 million and an operating profit margin of approximately 7%. Terms of the deal were not disclosed. Bay Minette, AL-based Quincy Compressor was founded in the 1920s and designs and manufactures reciprocating compressors, rotary screw compressors and vacuum pumps, primarily under the Quincy brand. Quincy products are sold through a network of independent distributors and agents. More than three quarters of the company's sales are in the United States.

Atlas Copco said it plans to further develop the Quincy brand independently, in line with the group's current brand portfolio strat

"I am pleased that we have obtained the approvals in the United States and can now move forward to further develop the Quincy business," said Stephan Kuhn, president of Atlas Copco's Compressor Technique division.

Approvals from Chinese authorities are expected during the second quarter.

Field Test: Tube-Tech CL1 B Compressor/Limiter

Compressor/limiter became a coveted and trusted dynamics processor noted for its great sound in a wide range of applications. The CL1 B has its optical gain-changing element (LED- and light-dependent resistor) placed directly after the input transformer, so there's no added coloration or distortion from an audio input amplifier. Two solid-state, time-constant circuits (one fixed, the other for variable or manual) control the opto element's attack and release characteristics, followed by an all-tube signal path with up to 30 dBU of makeup gain. For external sidechain interconnection, the CL1 B uses two buses. Up to 10 units can be slaved for processing multitrack stems or surround material.

Enter the Virtual CL1 B

Previously offered only for TC Electronic PowerCore users, the Tube-Tech CL1 B virtual compressor/limiter plug-in is now available for Digidesign Pro Tools HD, HD Accel and VENUE systems. It's for TDM systems only (Mac or PC) and works at sample rates up to 96 kHz, precisely emulating the hardware unit in all ways — including its Fixed, Manual and Fix/Man modes.

Fixed mode has 1ms attack and 50ms release times. Manual mode has controls for continuously varying attack from 0.5 to 300 ms and release times from 50 ms to 10 seconds. The plug-in retains the hardware CL1 B's unique Fix/Man mode, which combines the attack/ release times of Fixed and Manual modes. Designed for program compression, Fix/Man's attack time is the same as Fixed as long as program peaks are longer than or equal to the attack time setting. For program peaks that are shorter than the attack time setting, the attack time can control or delay the onset of the release timing.

The Fix/Man mode results in a variable release with a constant fast attack (when fully counterclockwise). You'll have faster release times for short-duration peaks and longer release times for program peaks with longer expires.

You can vary the compression ratio from 1:1 to 10:1, and a large VU displays input, output and gain-reduction levels. The stereo plug-in version allows you to select the channel — left, right or a combination — that drives gain reduction.

Taking it for a Drive

After installing CL1 B into my Pro Tools HD3 Accel rig and authorising via iLok, I replaced my usual compressor choices for vocals, bass, drums and backing vocal stems. At 48kHz sample rates, one mono or stereo instance uses 50 percent of a chip while 96kHz sessions use an entire chip per instance. Its 28 presets offer many good starting points. You can toggle between A/B compressor settings. "A" lights up as soon as you tweak any preset.

My first session had a difficult male vocal track that eluded previous processing, with no middle ground between oversquashed and loose control. The CL1 B's Male Vocal preset came ready to work with fixed attack and release times (those knobs were grayed out), so I changed to Manual mode and lengthened the attack time to regain some lost brilliance and experiment with different ratio settings. In Manual, I had to reset the threshold, which was too bad because I liked the preset value. I wish that the current values of attack, release, threshold, ratio and makeup gain were indicated in boxes in which you could type values.

As with the hardware unit, the singer sounded full and had a very high fidelity — when peaks resulted in 10 dB or more of gain reduction, the vocal didn't sound distorted or extremely dulled out.

The next track was a Fender bass. Like its hardware cousin, the CL1 B plug-in may be the best compressor for bass guitar because it's clean, yet offers a tube warmth that imparts harmonic richness. I used Fix/Man mode so that the attack time turned into a kind of hold function. This gave me the best of both worlds: a long release time for the legato bits — whole notes in the verses — and then faster release times for the straight eighth notes in the choruses. I set the attack control to the full-clockwise position and release to about 2 o'clock. The verse "footballs" were held to a constant level while the eighth notes in the chorus chugged along evenly.

On kick drums, the CL1 B tightened the low frequencies considerably. I started with the Kick Drum preset and got more punch and aggressive attack with little sacrifice of the drum's acoustical size. The CL1 B will do its own unique version of "crazy drum limiter" that's cleaner (with fewer distortion artifacts) than other plugs. Set the ratio to 10:1, turn the threshold down to about -40 dB and set a fast attack time, but play around with the attack time and the release time depending on a song's tempo and the drummer's style.

I used my ears more than the CL1 B's metring — what looks like a lot of gain reduction on the metre doesn't necessarily sound like it. It's easy to overload the plug-in, with red LEDs lighting up long before the GUI's output metre reaches mid-scale.

But my ears kept confirming that the CL1 B sounds great — it was excellent on every source I tried. And like the hardware unit, its smooth operation and fat sound never disappoint. The CL1 B is first-call and first-class for vocals, bass and guitars — but use it on any source. At $995, it's a winner.

Ingersoll Rand Launches Garage Mate Compact Compressor

This is the ideal compressor for professional and home use where users need high performance, but space is at a premium," said Dipen Shah, global portfolio manager for small reciprocating compressors at Ingersoll Rand. "The Garage Mate delivers all the power our customers need for their toughest applications, while conserving room for other equipment."

Designed for professionals and serious do-it-yourselfers, the versatile Garage Mate compact compressor delivers maximum air for impact guns, ratchets, grinders, drills, nail guns, paint sprayers, sanders and more.

The Garage Mate P1.5U-A9 is a 2-horsepower, portable single-stage compressor that delivers 5.2 cfm of air at 90 psi and features an easily accessible drain valve. With durable cast-iron cylinder and frame construction, this compressor is built to last and to deliver plenty of power for 100 percent continuous duty on the toughest applications. The Garage Mate is also ideal for users who want a compressor that can meet their demands today and continue to meet expectations as their skills and requirements grow.

In addition, a belt drive makes the Garage Mate quiet. At just 76 dBA, it's a compressor that's easy on the ears in both residential and commercial settings.

Yen Sun Technology Corporation, commonly known as Y.S. Tech, make computer fans. They make a lot of computer fans. And all of their products, up until now, have been pretty similar to the products from other computer fan manufacturers.

Different fan manufacturers make fans with somewhat different blade profiles and somewhat different motor designs and various other points of differentiation, but all normal computer fans are similar in their basic architecture. Frame, impeller, ball or sleeve bearings, brushless motor in the middle of the impeller.

Figure: Well, here's something a little different.

Y.S. Tech's new Tip-Magnetic Driving Fan moves the motor out of the hub of the fan, and puts it around the edge. The impeller blades are surrounded by a ring studded with 12 magnets, which are acted upon by four coils that're located at the corners of the fan's frame.

Here's the TMD fan with its lid removed. You can see the four coils, and the little circuit board that controls them.

In essence, this motor design is the same as that of any other brushless fan. Stationary coils, rotating magnets, Hall effect sensors that let the controller know how the rotor's oriented, so it can energise the coils at the right time.

But instead of bunching the whole assembly up inside the fan hub, with the coils sitting on top of the controller board and the magnets surrounding them, everything's been spaced out.

Y.S. Tech have a "technical issues" page for the TMD fans, which explains that this design yields simpler coil windings, more efficient magnetic field structure, assembly as a simple gomphosis (your Vocabulary Word Of The Day), and lower acoustic noise, thanks to the ring-surrounded blades. On their Application Example page, Y.S Tech further explain that the smaller hub creates less of a "wind shadow" in the middle of the fan. On a CPU cooler, that's where you want more air flow, because the middle of the fan's probably right over the contact point where the heat enters the heat sink.

Which is all very well. The question is whether these fans actually work any better than the ordinary sort.

And the answer, right now, for many people, is no. You don't even have to test the things to know that. Because the first models of TMD fan are 70mm units - bigger than the 60mm fans used on most CPU coolers, but smaller than the 80mm ones used for most case-cooling purposes.

70mm fans are just right for various P4 coolers, and for the few 70mm-square coolers for Socket A and Socket 370 processors. But most people won't be interested until Y.S. Tech release 60mm and 80mm TMD fans, which are promised in the near future.

Bitspower, however, sent me a Socket A/370 CPU cooler with a 70mm TMD fan on it. And, conveniently, two other coolers that use the exact same heat sink as the TMD-fan one, but have normal 80mm fans on them instead.

Figure: The TMD-fan cooler is the NP81DT58, which isn't commonly available yet, but can be purchased directly from Bitspower for $US45 (or 50 Euros), for which price you also get one of their new BP478R adapters, which lets you use a Socket A/370 cooler on a Socket 478 Pentium 4.

According to Bitspower, this price includes Fedex shipping "everywhere". Mawson Station residents are probably excepted, though. The NP81DT58 uses the "middle speed" 5800RPM PD1270155B-2F TMD fan; this fan's the fastest TMD that exists so far. The other options are "low speed" and "extra low speed", model numbers PD1270153B-2F and PD1270152B-2F, respectively. The low speed TMD will be used on Bitspower's upcoming NP81DT45.

Figure: The other two Bitspower coolers look identical to each other, and they both use the same 80-by-60mm NP81 heat sink as the TMD-fan unit. These ones have a different frame on top, though, to suit an 80mm fan. They're the "Silence" NP81DS30 and the "Super Silent" NP81DS20, with 3000 and 2000RPM fans respectively.

How do these fans compare with the TMD? Well, there's an Acrobat-format datasheet for the PD1270155B-2F TMD fan. If you don't have the current Chinese Traditional font pack for Acrobat, you'll need it to read all but the last couple of pages of the spec sheet.

The flow-versus-pressure graph at the end of the TMD datasheet shows a pretty smooth line from a maximum flow rate of almost 37 cubic feet per minute (CFM) at near-zero static pressure (this is the "free air" result), to a maximum static pressure of 0.2265 inches of water (in H_2O) at zero flow (this is the "sealed box" result). In metric, that's 1.04 cubic metres per minute and 5.755 millimetres of water, respectively.

The smoothness of the graph tells you that there's nothing peculiar going on; this isn't some odd fan design that gets tons of air flow in an unrealistic free air test, but can't do much in a real-world situation where there's some flow resistance to deal with.

For a 5800RPM fan, by the way, this one is pleasingly quiet. The spec sheet quotes a noise level of 39.6dB (A weighting) at one metre, and that was with the fan running at more than 5900RPM. There are plenty of 60mm fans around the 5500 to 6000RPM mark that're significantly louder than this; they're all taller units than the TMD fan, but the TMD does better than most of them for air flow, in the free air figures, anyway. Taller fans can generate more pressure, all things being equal, so the TMD is probably broadly even with these louder 60mm units, if you ignore near-field turbulence.

I don't have good numbers for the fans on the other two Bitspower coolers (and, in any case, the manufacturers promise on the pages for the coolers that "The Specification of Fan will be changed without notice"...), but Bitspower rate them at 47.8CFM for the 3000RPM one, and 30CFM for the 2000RPM one. Those are free air ratings of dubious veracity, but unless the numbers are very inflated, they still mean that the NP81DS30 should outperform the TMD-fan cooler, if the TMD fan's different air flow dynamics don't help it.

So I tested all three coolers, on the same rig I use for my big CPU cooler comparison here.

The DS20 managed 0.73°C/W. A perfectly all right result from a nice quiet cooler. The DS30 scored 0.66°C/W; again, about what I'd expect. And the DT58, with the TMD fan obviously shifting more air over the heat sink than the DS30's fan could manage (and obviously making more noise, too), clocked 0.62°C/W.

Which isn't particularly thrilling, but it's less than 20% behind the results of the very best Socket A/370 coolers I've tested. This is all you can expect from a large all-aluminium heat sink without a stupid deafening fan on it; the NP81's got fancy skived fins, but diminishing returns cut in ferociously as you make aluminium heat sinks bigger, no matter how spiffy the fins are.

So there would definitely appear to be something to Y.S. Tech's claims about the TMD fan. The thing isn't an incredible revolution in air-shifting technology, but it does work well; compared with the average low-profile 70mm fan, the TMD's a monster, without being horribly loud.

Performance isn't everything, of course. Price is an issue. And so's coolness. TMD fans look likely to be winners in these areas, too.

How much?

Y.S. Tech have a where-to-buy-TMD-fans page here which, in the great tradition of where-to-buys from Taiwanese companies, leads to various sites that don't list prices, have buy-it-here links that don't work, or don't list the product in question at all.

One of 'em worked, though. It would appear that European buyers will be able to get themselves a TMD fan for 15.50 Euros (about $US14.20, as I write this) plus another 5 Euros for shipping, from Elektrosil Systems' tmdfan.com site. That site says the fans will be available from the 15th of May.

That's a good price. It's less than you'll pay for the super-high-power fans that many overclockers stick on their CPU coolers.

The medium power TMD doesn't have the same air moving ability as an above-six-watt screamer of conventional design, but its more effective airflow seems likely to help, as should the new TMD models on the way.

Atomic batteries to power

And then there's the coolness issue. Will TMD fans impress your fellow computer-tweakers?

Yes. Yes they will.

The TMD fan's impeller and motor, together, make it pleasingly weird for people who're used to normal computer fans.

The impeller looks nifty, of course, but its design also means it's considerably heavier than usual, and also has a high moment of inertia. Most of its mass is around the edge, a long way from the pivot point. Normal fans have their heavy magnets in the middle; this one has them around the rim.

What this means is that the TMD fan takes an impressively long time to spin up to full speed. More than nine seconds, in fact. Most computer fans this size spin up in around three seconds.

The TMD fan also sounds interesting when it's spinning up. The motor controller creates a distinctive electronic squeal on startup, which settles down to a relatively quiet whine that's swamped by the noise of the blades as the fan gets to speed.

I've made a 387 kilobyte MP3 of the TMD fan spinning up and then being turned off. The mike's quite close to the fan for this recording, which means it sounds rather more like a jet car than it does in real life, at least if you don't have your ear close enough to the fan that it's trying to eat your hair. But it conveys the general impression.

I can only imagine that the 80mm TMDs will take even longer to spin up, and sound even more entertaining when doing it. Which'll make them excellent candidates for anybody who's been looking for the perfect 16 fans from which to assemble a whole side panel for his or her computer case. The heavy rim, by the way, also makes the TMD fan highly gyroscopic. When it's running, it doesn't want to change orientation. All fast fans are somewhat gyroscopic, but this one's quite powerfully so. For computer applications this isn't likely to make any difference to anything, but it's quite fun if you're just playing with the bare fan.

The TMD fan's stall behaviour is peculiar, too. Normal brushless-motor computer fans don't die if you stop them from turning while they're powered. They just sit there, drawing maybe three or four times more current than usual. Brush motors draw more power and deliver more torque the slower they're turning, and can quite rapidly cause physical or electrical havoc if they're jammed solid, but brushless motors are smarter about it.

The TMD fan's smarter still. If you stall it - by sticking your finger in it as you power it up, for instance - it tries to drive for a moment at quite low torque. Your finger's in no danger at all. Then it goes silent, for two to six seconds, according to the spec sheet. It draws no current while it's silent. Then it tries its motor again, and goes quiet again if it fails to turn. The TMD fan I played with pulsed steadily, once every 1.4 seconds or so.

This means that a jammed TMD fan definitely won't overheat. Whatever it's meant to be cooling might be toast, of course, but the fan at least is in no danger, and the rhythmic squeak as it tries to spin may attract your attention before any damage is done. Regular fans are silent when they're stalled.

The TMD fan, by the way, has a normal three-wire connection, so you can hook it up to a motherboard fan header and monitor its speed in the usual way, if your motherboard supports that. Most do, these days. Fan speed monitoring lets you run software that'll sound an alarm or shut down if a fan stops turning, which is better than some other systems for dealing with dud fans.

Stupid Engineering Tricks

While I was playing with the TMD fan, it struck me that for really optimal performance, a computer fan's blades need to not only be driven from the rim, but also supported there. The TMD fan gets rid of the bulky motor assembly in the middle, but the fan's spindle and bearings are still central. And so there still has to be a little boss in the middle of the fan and, more importantly, supporting arms coming in from the frame. The arms and the spindle base need to be reasonably thick, to give them some strength in case something whacks the impeller. The TMD fan's three spiral arms are thinner than usual - partly because none of them need to provide a wire-route channel - but they're still right next to the blades, and must therefore still cause considerable flow disturbances.

If the impeller were supported by a bearing around its edge - if the whole impeller were built as an extension of the inner ring of some kind of ball or roller bearing - then there'd be no need for the spider of support arms and the central bearing assembly. Thought I.

It then struck me that, unfortunately, this ain't likely to happen. Not at a reasonable price, anyway. Because stuff would have to move too fast.

The centre support bearing of the 5800RPM TMD fan has to spin at 5800RPM. Duh.

The fan's impeller ring, however, has a diametre of about 70mm (the whole fan's a bit big for a "70mm" unit). So its circumference is about 220mm, and it spins 5800-odd times a minute, or about 97 times per second. So the perimetre's moving at about 21.3 metres per second - around 77 kilometres per hour.

The inner ring of the two centre ball bearings is less than 5mm in diametre; it's unlikely to be moving faster than about 1.5 metres per second. Maybe 5.4 kilometres per hour, tops.

Now, the inner bearings are teeny, but unless the balls in them are really amazingly small, they won't be taking a great deal of punishment.

Replace the little bearings with a big outer ring and even if it used big beefy balls, they'd still be ripping around the bearing at a hilarious rate of knots. If you tried to avoid making a large precision bearing by putting a rail arrangement around the edge of the fan and supporting it with three or four smaller bearings (which would leave gaps between the supports for the drive coils, which would otherwise have some trouble getting at the rim magnets), then the little edge bearings would be spinning even faster. A 10mm bearing in contact with the edge would have to spin at more than 40,000RPM. Centre bearings are, therefore, here to stay. Well, unless and until someone comes up with an air-pressure or magnetic or similar clever-clogs non-contact bearing design that'll work in a moulded plastic fan that sells for less than the price of a Rolex.

Overall

The current TMD fans are genuinely superior to similarly specified conventional fans. But they're the wrong size for most applications. I'll be interested to see the 60 and 80mm ones when they come along.

If you simply must have the weird spool-up sound of a TMD fan in your life right now, then I suppose you're going to have to buy one or more 70mm units as soon as you can.

Chapter 7

Solutions for Industrial Fans

Thousands of customers have trusted us to fabricate & repair their air handling centrifugal fans, fan rotors, and complicated structural support. With production facilities in Alabama, we offer turn-key industrial fan solutions from design and fabrication to installation and commissioning throughout the United States, Canada, and Latin America.

New Centrifugal Fan Design

A centrifugal fan has a moving impeller that consists of a central shaft with a positioned set of blades. Centrifugal fans blow air at right angles to the intake of the fan, and spin the air outwards to the outlet (by deflection and centrifugal force). The impeller rotates, causing air to enter the fan near the shaft and move perpendicularly from the shaft to the opening in the scroll-shaped fan casing. ProcessBarron's centrifugal fans include:

ProcessBarron's centrifugal fan design evaluation includes assessment of temperature, abrasion, corrosion, stress, and harmonics to generate the most efficient, economical, and productive centrifugal fan design for your application. All new industrial fans are performance rated and calculated in accordance with AMCA (Air Movement and Control Association) Standard 210-07 Laboratory Methods of Testing Fans for Aerodynamic Performance Rating/ASHRAE Standard 51. Additional centrifugal fan construction services include:

- Finite element analysis for stress analysis of rotors using ANSYS software.
- Centrifugal fan drawings provided in a CAD format (AutoCad) for easy import into existing customer electronic drawings.
- ANSYS CFX - Fluent computational fluid dynamics to analyse and accurately predict fan performance. This computer

simulation technique is particularly useful when upgrading a new rotor into an existing housing.

- Modal analysis to determine natural frequencies of the as-built rotor to prevent resonance problems and premature failures.
- All rotating assemblies welded to AWS (American Welding Society) Standard D14.6 "The Specification for Welding of Rotating Elements of Equipment."
- Balance all rotating assemblies to ANSI (American National Standards Institute) Standard S2.19, Grade 6.3.

ProcessBarron also offers complete fan repair and fan testing for any size or type centrifugal fan impeller, at our facilities or on-site in the field. We are familiar with erosion, corrosion, material fatigue cracking, and a host of operational problems that can lead to catastrophic failure.

Our industrial fan repair, rebuild, and upgrade services can provide years of additional fan service. Learn more about our full range of field and technical services, or call our 24-hour hotline (800-226-3267) for emergency repair and services.

Centrifugal Fans - Impeller Designs

Howden supplies centrifugal fans with five basic blade shapes, and a number of impeller configurations. The final choice of impeller design will depend on the aerodynamic duty and the operating conditions.

Our experienced engineers work with customers to determine the optimum solution for a particular application.

Backward inclined, backward curved blades

A highly efficient design of impeller. Its inherent strength means that in addition to handling air, it can be used with gases containing moderate amounts of erosive particles and at high temperatures. It can carry liners and hard surfacing, if required.

Backward inclined, flat plate blades

An efficient and strong shape, this design is a cost-effective alternative to the backward curved design but with flat plate blades instead of curved. This results in a slightly lower efficiency, compensated by the easier fitting of liners.

Backward inclined, forward curved blades

Also termed radial tipped blades, this design reduces the potential for dust build-up on the underside of the blades in applications with moderate amounts of gas-borne dust.

Backward inclined, aerofoil blades

Aerofoil bladed impellers are most commonly used in applications handling large volumes at low pressures. This highly efficient design of fan is used in clean air or with gases containing small amounts of erosive particles.

Radial blades

Forming a rotor which is essentially a large paddle wheel, this design results in a relatively inefficient fan with a power consumption higher than that using the much more common backward inclined blade.

Its inherent mechanical strength and resistance to wear mean it is generally used when high quantities of abrasive dust are present in the gas stream, or when very high gas temperatures are expected.

Impeller configurations Single or double inlet

A double inlet impeller consists of two single inlet impellers back to back on the shaft sharing a common backplate, providing almost double the flow of a single inlet fan. The selection of a double inlet fan results in a smaller diametre, faster (and perhaps cheaper) machine compared to the single inlet alternative.

Single or multiple stage

A multistage fan consists of two or more impellers mounted on the same shaft. The air or gas flow passes through each one in turn, with a consequent increase in pressure.

How to Design a Centrifugal Fan

A centrifugal fan is a rotating piece of equipment that has vanes or blades contained in a housing or casing. The spinning vanes produce air flows with high volumes, but low pressures. Typically, an electric motor drives the vanes and as the vanes rotate, the air is pushed against the outer walls of the housing and exit at a right angle to the intake. The action of the air being thrown against the walls provides the centrifugal force that drives the air stream.

Instructions

Determine the maximum volume of air needed for the process. For instance, assume the air volume needed is cubic feet per minute (300 feet per second).

Determine the size of the outlet duct based on air velocity. For example, good engineering practice specifies air velocities to be between 70 feet to 100 feet per second. For this application, assume a velocity of 90 feet per second.

The cross sectional area of the duct will be the volumetric flow rate divided by the velocity or 300/90 which is 3.3 square feet. For a square duct, each side can be approximately 2 feet and the outlet of the fan should be 2 feet by 2 feet.

Determine the size of the electric motor to produce the volume required. Fan manufacturers can supply a value called the displacement of air per revolution ("V"). This means for each rotation of the vanes, a specific volume of air is pushed out. A typical rotational speed for an electric motor is 3,500 revolutions per minute (rpm) with an efficiency of 70 percent ("n"). Assume the air displacement is 50 cubic feet per revolution. The equation to determine the motor horsepower (hp) is hp = V x rpm/396,000 x n. This calculates to be 50 x 3,500/ 396,000 x 0.70 or 0.63 horsepower. Therefore, a 3/4 horsepower motor can be used for this application.

How to Build a Centrifugal Fan

A centrifugal fan moves gases such as air, around an environment. Also known as blowers, tornado fans or squirrel-cage fans, these fans generally rotate around a driveshaft. The gas being driven is drawn in through the side of the fan's wheel and is turned at a right angle before being expelled. Because of the centrifugal forces involved with turning the wheel, the gas is moving with considerably more force when it leaves the machine. For high school students looking to explore these forces in a science fair project, it is fairly easy to build a centrifugal fan and use it for testing its effects.

Use the drill to make a 5/16-inch hole in the middle of the fan plate. Slide the bolt down through the collar for the shaft and the plate. Use a nut on the other end to secure it.

Weld the collar down to the plate, leaving a gap for the set screw for later. Loosen the nut and take it and the bolt out.

Add a fan blade to each slot. The outside edge of each blade should rest evenly with the edge of the plate. Weld the blades into place by joining them to the plate at two or three different spots. However many welds you use, spread them out evenly.

Put four bolts into the housing plate for the bottom. Tighten with nuts and then weld the bolt heads so that they won't move. Slide the side strip into the tabs that you will find on the housing plate. Weld each tab into place before you insert the next one. When you're done, turn the plate over and weld the tabs into place where they stick through the slots as well.

Repeat Step 4 for the top plate, using the same side strip attached to the bottom housing. Slide the nipple onto the outlet hole and weld in place, all the way around the nipple. Place the housing onto the motor and slide the fan inside.

How to Increase Pressure on a Centrifugal Fan

The pressure generated by a centrifugal fan depends on a number of factors. If the increase in pressure is needed for an extended period of time, it is usually easier to purchase a new fan with different characteristics than to modify the pressure output of the existing fan. However, there are a few tricks you can try to change the pressure for short periods.

It should be noted that excessive pressure above the manufacturer's rating can damage the fan. Check the rotation speed of the motor that powers the fan. If the fan and motor are directly coupled, replace the motor with a larger motor with a slightly higher speed than the existing motor.

The motor will need to have at least the same amount of torque as the original motor. If the motor is connected to the fan with a belt-and-pulley system, go to Step 2.

Remove the fan pulley. The fan pulley is usually larger than the motor drive pulley. Replace the fan pulley with a slightly smaller pulley, then reinstall the pulley with a new, smaller belt. This setup will increase the speed of the fan and the pressure it generates.

Reduce the size of the fan's outlet. This forces the air exiting the fan to speed up. As the opening deceases, the air pressure increases. You can do this by installing an outlet reducer that can be bolted onto the front of the fan. Smaller fans can use PVC pipe reducers.

Centrifugal vs. Vane-axial Fans

The choice of forward-curved centrifugal or backward-inclined centrifugal or airfoil centrifugal fans versus vane-axial fans will affect the energy efficiency. Standard centrifugal fan systems are typically very inefficient (50 percent to 60 percent), and their actual installed efficiency is usually much worse than the manufacturer's ratings. Efficiencies as low as 30 to 40 percent have been measured in the field. It is recommended that high-efficiency (80 percent to 90 percent) axial airfoil fans be considered with direct drive and static regain devices. However, flexibility of adjustments will be compromised.

Fan Size vs. Speed

Fan size influences horsepower requirements, with larger fans tending to have higher efficiencies than smaller fans. Larger, low-speed fans also have reduced drive losses and lower maintenance costs.

Sound Power Levels

If a vane-axial fan is selected, louder fan sound may require attenuation, resulting in system pressure drops. Even though some centrifugal fans have lower efficiencies, they can operate at lower sound levels, eliminating energy-wasting pressure drops

System Dynamics

The design engineer should investigate various options for the prescribed laboratory configuration and examine the total system dynamics, not just one component such as system static pressure. When the entire system is evaluated, unexpected results may occur. Kruse (1991) points out that, "Until all other components of the air-side system, including ductwork and filters, have been designed, the fan(s) cannot be selected."

This is contrary to typical design practice. Engineers should consider the brake horsepower required for not only the design condition, but also all other conditions that the fan may experience, such as part-load, start-up load and full RPM load. Also, the number of fans used in a pressurised plenum design influence system redundancy and flexibility as well as total energy use

Selection and Optimum BHP

Kruse (1991) presents and analyses "published fan performance data for single-width, single-inlet, air-foil style plug fans at 20,000 CFM and 2.5 inches static pressure":

Size	BHP	RPM
33	15.6	1,200
36	13.8	960
40	12.3	767
44	12.1	653
49	12.6	586
54	13.8	524

These data indicate two common trends. First, as fans get larger, they perform the same work at lower RPM's. Second, there is an optimum BHP selection with poorer selections for larger and smaller fans. Based purely on the above data, a size 44 fan would be the optimum selection.

The worst fan would consume 29% more energy. Two factors, however, might lead to the selection of the size 40 fan, even though it consumes 1.6% more energy. First, the size 44 might simply not fit in the available space. Second, if both are driven by 900 RPM direct drive motors with variable-frequency drives (VFD's), the size 40 fan would draw 20 horsepower at 900 RPM while the size 44 fan would draw 30 horsepower at 900 RPM. This condition could easily occur if the VFD went into a bypass mode. Providing a 20 HP motor for the size 40 fan makes more sense than buying a 30 HP motor for the slightly more efficient size 44 fan, considering the costs of the fan, the motor, the wiring, the VFD and the lower efficiency of a 30 HP motor operating at 12.1 horsepower.

Fan Selection Evaluation

In evaluating any fan system and the corresponding system effects, there are two basic areas of consideration: those elements that are constant from one selection to the next and those that are variable based upon which fan is utilised.

Constant Elements

- Filter Pressure Drops—Based on constant flow rates, filter pressure drops remain the same between alternate fans.
- Coil Pressure Drops—Based on constant flow rates, coil pressure drops remain the same between alternate fans.
- Airflow—Airflow rates are maintained as a constant between alternate fans.

- External Static Pressure—The resistance to airflow external to the unit discharge and prior to the return air inlet is considered a constant.
- Inlet/Outlet Sound Power Specification—Inlet and outlet sound powers are a constant, given for compatibility to room sound level specifications. Although levels might vary from selection to selection, the specification will remain constant.

Variable Elements

Total Pressure – In all systems where air is moved, a fan must develop pressure to overcome all friction and dynamic losses within that system. The pressure that the fan generates is total pressure, which is the sum of two components, static pressure and velocity pressure. Static pressure is that portion of the air pressure, which exists by virtue of the degree of compression only. If expressed as gauge pressure, it may be negative or positive. Velocity pressure is that portion of air pressure, which exists by virtue of the rate of motion only. It is always positive. Total pressure is the air pressure, which exists by virtue of the degree of compression and the rate of motion. It is the algebraic sum of the velocity pressure and the static pressure at a point.

Noise Control – Inlet and outlet sound power specifications are constant system criteria. However, to meet stringent clean room sound level specifications dependent upon the fan selected, based upon physical constraints, varying amounts of noise control will be required within the system. Noise control elements often add static load to the fan, which results in increased fan noise. This increased fan noise must be considered in the choice of the control element.

Fan Efficiency – The efficiency of a fan is an important element in the selection process and an element that will vary from one fan selection to another and from manufacturer to manufacturer. With many clean room applications where large airflow rates are involved, optimising system efficiency can result in substantial operating cost savings. Equally significant, an improper selection can yield large penalties in operational costs. Concentrating solely on one aspect of the system will often result in a fan system's failure to achieve its highest overall efficiency. Often the critical error is concentrating solely on fan efficiency. The important thing to recognise is that catalogue data and efficiency ratings are developed under "ideal" test conditions, which are rarely duplicated in clean room configurations.

As a result, the actual field conditions must be carefully evaluated and an analysis made of their effect on fan performance with corrections being made both from factory-supplied data and other sources, such as AMCA publications, personal field experience, and private lab testing.

System Effect – The term "system effect" is a subjective term, intended to ask how the fan's performance is affected by external conditions. Critical examination of the inlet and discharge conditions relating to the fan is appropriate whether utilising an axial or centrifugal device. The application of correction factors relating to fan performance is necessary where field conditions vary from test conditions. Application of these factors will essentially ensure that full air delivery to the system will be accomplished.

System Effect

Specifying a high efficiency fan out of a catalogue is not enough. Careful design requires looking at all aspects of the fan-motor-housing system....[In] an example...an otherwise efficient direct-drive axial fan...had a motor mounting scheme that covered up nearly a third of the fan opening. Likewise, it is not enough to just specify fan efficiency at a single point, it is also important to specify a fan with a large "bulls-eye"; i.e., where the peak efficiency can be maintained over a wide range of static pressure and fan flow rate....efficiency plots from one manufacturer of virtually identical models [can be quite different], one with a very small efficiency "bulls-eye" and one with a very large "bulls-eye" for an otherwise identical "80% efficient" fan.

Variable Frequency Drive (VFD) Application

Volume control can affect the efficiency of the air system. Belt drives are common and work well for non-clean purposes such as offices, but the particles they shed preclude their use in clean rooms. Dampers and inlet vanes can throttle a fan to the desired operating point with some energy savings. One plug fan manufacturer has a patented device which goes one step better by effectively reducing the fan width. With vane-axial fans, the blade pitch can be changed, either while the fan is running (low RPM fans only) or while the fan is off. The RPM has to do with the mechanism involved and not the skill of the mechanic. VFD's can be used on any style of fan and are very efficient. Depending on the style of fan being used and the flexibility desired, blade pitch adjustment or VFD control will be the most efficient.

Fan speed is controlled by the lowest reading of multiple-duct static pressure sensors. Each fan receives the same control signal, thereby providing nearly equal loading. This loading is dependent upon manifold design; however, there is an equalizing effect when any imbalance occurs. The system operates at a constant volume, as determined by the flow setting of each constant-volume terminal unit. Even though the system is constant volume, VFD control is provided to account for variable filter loading, future capacity changes, and three- or four-unit operation without affecting system balance. It should be noted that the constant-volume design was predicated on the fact that the majority of laboratory exhaust is through process equipment, customer-fabricated hood/enclosures, laminar flow fume hoods and gas cabinets, etc., all of which are not adaptable to current VAV control technology.

Drive and Motor Efficiency

The choice of drive mechanism and motor type will also yield considerable energy-savings potential. Additional energy savings of 2 to 5 W/ft2 (21.6 to 54 W/m2) may be achieved with the substitution of high-efficiency motors for standard-efficiency motors. The choice to use fans with motors outside the primary air stream will save 1.0 to 1.5 w/ft2. (10.7 to 16.15 W/m2) due to motor inefficiency, but removal of the motor from the air stream by using belt-driven fans with lower mechanical efficiencies will reduce the energy savings. High-efficiency motors have an inefficiency of 6% to 8%, whereas belt-drive systems have inefficiencies of 4% to 6%. The use of special direct-drive fans with motors outside the air stream (bifurcated fan housings) may result in higher system effects and lower total efficiency

Synchronous Belts

Synchronous belts are stretch-free and can achieve higher efficiency than V-belts due to lower hysteresis, lower friction losses, and truly synchronous operation (no slip; no creep). The lower flexing losses result from the small thickness of the belt between the teeth. The absence of speed losses (slip and creep) is due to the positive mating of the teeth of the belt with the teeth of the sprockets. Because of these lower losses the synchronous belts have not only higher peak efficiency than V-belts, but also the efficiency curve is flatter.

Synchronous belts require toothed pulleys (sprockets), which are more expensive than V-belt pulleys. In a retrofit application, the additional investment required is the cost of the sprockets, plus the extra cost of the belts and of the installation. In the cost-effectiveness calculations it is assumed that those extra costs fall in the range $8-16/hp. This price variation is strongly dependent on the rating of the belt drive, with the large drives requiring a smaller price premium per horsepower. The sprockets typically cost 1.5-2 times the cost of belts (the second and third purchased at future times and thus discounted) and one set of sprockets. Additionally, it is assumed that there is an average 5% efficiency improvement and that the belt will last 24,000 operating hours. Synchronous belts provide slip-free synchronous operation, have a long lifetime, and can operate in wet and oily environments. Their operation can be noisy at high speed and their main weakness is the lack of capability to absorb shock loads.

Scroll Housing Effects

Eidam (1988) recommends against fans with scrolls. "One would prefer to have a fan without a scroll which could just be bolted into the housing of the cleanroom equipment without any additional connections on the inlet and/or pressure side. The scroll housing is not desirable because the proportion of dynamic pressure maintained in the discharge of our fan is of little use to the cleanroom equipment."

Manufacturers' Data

When comparing fans, remember that fans are mass produced and that fan performance data (curves) has [sic] evolved over time without the benefit of a factory test on each and every fan ever produced. This means that one manufacturer may publish different data than another manufacturer for the theoretically same fan. Although some of the difference could possibly be due to minor design differences between the two fans, it could also be due to a series of statistical interpolations over a period of time.

Coils

The selection of hot and chilled water coils will have a substantial impact on the fan energy use.

Thin Coil Design

Traditional AHU design specifies coil sizes assuming a face velocity of between 400 and 500 feet per minute. A new design technique called low face velocity, high coolant velocity or LFV/HCV has been researched at the University of Adelaide, Australia. This technique uses a "thin" coil design that is roughly half the number of tubes in depth as in conventional designs but double the coil face area. The net result is a face velocity in the range of 150 to 200 feet per minute (FPM) with much higher heat transfer efficiency and lower pressure drop than in conventional designs. Because the coil's pressure loss is proportional to the velocity at a square rate, face velocity reduction can result in pressure drops of one-fourth or less compared to the equivalent, traditionally designed coil.

Preheat Coils

A preheat coil is commonly used to control condensation inside the HVAC system for laboratories that use 100 percent outside air or when the outside air temperature falls below freezing. If a heating coil is used downstream, the preheat coil should become inactive to save energy when outdoor temperatures reach 45 degrees F. Preheat coils are also used to warm the outside air stream, assuring better air stream mixing and providing free humidification

Evaporative Cooling

Evaporative cooling systems have been used to condition facilities since before refrigerated cooling was invented. There are two types of evaporative cooling techniques: indirect and direct. The indirect approach removes sensible energy only from the air stream with an exchanger that has the air stream flow on one side of a plate and water on the other. The direct approach removes energy from the air stream adiabatically by adding moisture to the air stream. These approaches can be combined to provide cooling and humidification that displaces energy from a conventional compressor-type HVAC system. Annual energy savings of more than 15 percent for a one-stage, direct system and over 38 percent for a two-stage indirect/direct system can be realised when compared to a conventional HVAC system. Brown (1993) says, "The use of direct evaporative cooling to humidify increases the opportunity to recover heat by 36 percent and reduce energy consumption during the heating season." One- to two-year paybacks are possible. The justification of evaporative cooling should not be based solely on the ability to replace mechanical cooling but should be considered as an opportunity to displace mechanical cooling/ humidification energy.

Evaporative cooling should always be considered for all make-up air systems, regardless of geographic region, as noted in the descriptions of the approaches below:

Indirect

Indirect evaporative cooling is applicable when there is a large process equipment-cooling load, the HVAC system operates on a continuous basis, and a direct evaporative cooling opportunity exists. [Brown, 1993]

Direct

Direct evaporative cooling can be justified on the basis of eliminating the need for a steam humidification system, enhancing economizer operation, reducing humidification energy used in conjunction with heat recovery, and controlling the temperature of a space independent of controlling its humidity. Direct evaporative cooling for both make-up air and recirculation-type systems should always be considered when the outdoor wet-bulb temperature is 53°F (11.7°C) or less for 3,500 or more hours per year. Direct evaporative cooling of air conditioning condensers, which are normally simply air-cooled, can reduce electrical energy and demand consumption by 20 to 40 percent.

Two-stage

Typically, a two-stage evaporative cooling system comprises a direct evaporative cooler downstream of an indirect evaporative cooler. The two coolers can be combined in a myriad of arrangements to increase energy efficiency

Evaporative Cooler Configuration

Brown (1992) describes the use of evaporative cooling in a university life sciences laboratory in Arizona.

The indirect precooler is comprised of a standard finned-tube heat exchanger connected with piping to a cooling tower. Water cooled in the tower is circulated through the coil to precool outdoor air. Accordingly, mechanical cooling is displaced by indirect atmospheric thermodynamic exchange. (All evaporative cooling components are commercially available and can be competitively bid.)

Basis of Analysis

The baseline analysis is predicated on a conventional two-fan dual-duct HVAC system with steam humidifier (hereafter referred to as the conventional system). The actual system is identical to the conventional system except that a direct evaporative cooler is located downstream of the cooling coil in the cold deck, replacing the steam humidifier. In addition, the preheat coil is replaced with a preheat/precooling coil, which is connected via piping to cooling tower. A plate-and-frame heat exchanger in the piping loop provides a means to heat water during the winter.

Brown (1990) analyses the effectiveness of indirect evaporative cooling and concludes that:

Indirect evaporative precooling will only be effective where the difference between outdoor dry-bulb and wet-bulb remains high (13°F (7.2°C) or more) for the majority of the cooling hours. The concept is not applicable when the difference between dry-bulb and wet-bulb is small (12°F (6.7°C) or less) for the majority of the cooling hours. Thus, this energy conservation measure will be limited to dry climates.

Humidity Control

Humidification can be provided in one of several forms. The need for clean air which is free from particles and ions usually eliminates humidifiers which use untreated water, treated water or steam from treated boiler systems. This leaves steam and evaporative styles which use deionized water or reverse-osmosis reject water. The analysis actually hinges around two issues; the cost of the water and the cost of the energy to convert it into a usable form. Regarding the cost of the water, if the building has a [Deionized Water] D.I. production system which uses a Reverse Osmosis [R.O.] unit, use the R.O. rejected water because it is essentially free. Otherwise, use the same source of D.I. water as the rest of the project, and pay the price. Regarding the energy required to use the water, the most expensive is usually electric energy. Each analysis must be site specific, of course. Packaged humidifiers can use electricity to convert D.I. water into steam. These may be your only choice for small projects, but they will consume lots of energy on large designs. Packaged humidifiers can also use steam as an energy source to convert D.I. water into steam. This is usually more cost effective since the plant steam will have usually been made by using a lower-cost fossil fuel instead of electricity. Steam generators can produce D.I. steam in a similar fashion, but on a larger scale. Evaporative humidifiers are the best economic choice.... This is especially true with tight humidity tolerances and in dry climates where an adiabatic "free cooling" effect can be obtained. Altogether, the best combination is evaporative humidification fed by R. O. rejected water. Mumma (1997) provides the following advise regarding energy-efficient ultrasonic humidity control.

The ultrasonic humidifier is a more energy-efficient means of humidity control than steam humidification... Ultrasonic humidification requires a different approach when sizing the heating coil, which must not only be able to temper the mixed air but must also be able to provide sufficient heat to overcome the evaporative cooling effect of the humidifier (i.e., humidification energy). In steam humidification systems, the heating coil is not required to provide any of the humidification energy. Furthermore, the location of the SAT control sensor must be downstream of the ultrasonic humidifier. In addition, hunting is a situation that can frequently occur with ultrasonic humidification due to the multi-step nature of the device. The same hunting is not as likely with steam humidification if the steam control is fully modulating..

Brown (1993) identifies typical humidity ranges for laboratories.

For energy conservation reasons, the laboratory humidity...[is typically] allowed to vary from a maximum of 50% RH when outdoor wet-bulb is greater than 54°F (12.2°C) to a minimum 40% RH when outdoor air wet-bulb is 50°F (10°C) or less. Supply air dew-point control...[is] reset down accordingly. Face and bypass dampers...[are] controlled to bypass air around the evaporative cooler to maintain supply air dew point..

It is not possible to fully evaluate heat recovery potential and benefits without considering humidification. When humidity is added by steam, fuel-generated energy is always required. When humidity is provided with a direct evaporative cooler, all or a portion of the energy required to evaporate water can be supplied by recovered heat. While any supplemental energy required is fuel-generated, the source is not restricted to steam and the use of hot water is generally the preferred choice. As expected...the greatest requirement for humidification energy occurs in the colder/drier climates. Therefore, it is the author's opinion that a direct evaporative cooler be given highest consideration for humidification an energy conservation measure. ... In some climates, direct evaporative cooling has an added benefit by providing simultaneous cooling and humidification.

Economizers

Suhail (1989) points out that, "Laboratories, research facilities, clean rooms, and hospitals require large amounts of outside make-up air. These facilities can only benefit from economizers when the facility does not require 100 percent outside make-up air. Exhaust air from the facility may be added into this outside make-up air stream with an economizer to save energy. Inadequate mixing of the make-up air stream and the exhaust (return) air will defeat the function of economizer." Consequently, errors made by the mixed-air temperature sensor "can increase annual energy consumption by as much as 30 to 50 percent." In the same report, Suhail (1989) describes a study by Kao et al.of economizer operation, which found that "a sensor error of only 1°F wasted 10 percent of the cooling energy and 6 percent of the heating energy in the HVAC systems studied." To maximise the savings from economizer system, proper blending and sensing of the air streams must occur. Stratified, unblended joining of the return and outside air streams in HVAC economizer systems can result in varied building temperatures, uneven filter distribution, and, in the right climate, frozen coils

Mixing Box Efficiency

In the typical mixing box, outside air and return air pass through galvanized steel damper blades before entering the mixing chamber. Once in the chamber, the two air streams are expected to blend and then continue through the rest of the air handling system at a consistent temperature. However, standard mixing boxes can be limited in their ability to mix effectively due to poorly placed inlets, chamber leakage and dampers that freeze shut, allowing excessive amounts of outside air into a building. Properly proportioned parallel blade dampers are more effective in mixing air streams of different temperatures within the mixing box, but they are less effective than opposed blade dampers in controlling the air stream velocities.

The distance specified to mix air before it reaches the mixed-air temperature sensor may be inadequate in a typical box sized by the design engineer. In such cases the sensor will not detect the true average temperature of the mixed air; variances of more than 10 degrees F are not uncommon in the stratified air streams.

Air Blenders

The air blender is a static device that can be used to blend outside air and return air within a predictable standard deviation. Aerodynamically designed for manageable pressure drops, the standard blender is fabricated of 0.08 in.-thick aluminium. Blenders have also been used as a diffusion device on the outlet of a centrifugal fan to provide uniform velocity across cooling coils downstream. One aspect about air blenders...[to consider] is the length required to get mixing, which can add 5 to 8 feet to the unit....the length of ductwork that the blender requires for proper mixing... is a minimum distance equal to the diametre of the blender face.

Economizer Case Study

A second method of reducing the energy consumption of the operating rooms to incorporate variable air flow and the use of outside air according to occupancy and weather conditions was developed. The hospital requested and received permission from the New York State Department of Health to use the federal standard of 15 air changes per hour with 100% outside air when outside conditions made that selection more economical than the current state standard of 25 air changes per hour with 20% outside air. To enable the system to easily change back and forth from 25 to 15 air changes per hour, variable speed drives were installed on the supply and return fans. An energy management system was installed to control the heating and cooling coils, humidifiers, and variable speed drives. The system also monitors space conditions.

This system helps the HVAC systems run as efficiently as possible. The air flow rate is now determined by outside conditions; as a result, the system uses outside air with more frequency. The energy management system monitors outside air conditions during occupied hours and then determines whether the HVAC system should operate at 15 air changes per hour with 100% outside air or 25 air changes per hour with 20% outside air. When the outside temperature is between 35°F and 70°F, the system will operate according to the federal standard. When the outside temperature is below 35°F or above 70°F, the system will operate according to the New York State standard. During unoccupied hours, the systems operate at 15 air changes per hour with full return air.

Energy Efficiency and Energy Recovery

Energy recovery systems typically incorporate heat exchange equipment to reduce energy costs by extracting heat from the facility's exhaust air stream before it is vented outside. Energy recovery from the laboratory's exhaust should be considered when significant portions of operating hours are at ambient temperature of 50°F (10°C) and below. Another recoverable energy source is provided by chiller/DX condensers. Water cooled condensers can be piped to reject waste back into the labs HVAC system to provide reheat capacity, to augment run-around coil systems, and to dry regenerative heat wheels. When properly designed, these energy recovery systems can reduce installed HVAC system capacity by one-half; reduce operating energy from one-third to two-thirds, depending upon mode of operation; and have life-cycle cost paybacks from immediate to three years. The four major energy recovery systems include run-around coil systems, regenerative heat wheels, heat pipes, and fixed-plate exchangers.

Run-around Coil Systems

These systems can recover energy in other ways besides recapturing the heat in the exhaust air. For example, waste heat in the process cooling water from the laboratory equipment can be recovered. Water chiller waste heat can provide domestic hot water and space heating for laboratories and offices with a run-around coil system.

The following types of energy recovery systems require side-by-side exhaust and supply ducting.

Regenerative Heat Wheels

A regenerative heat wheel is a revolving disc filled with an air-permeable medium including a desiccant. When the air passes through the medium, heat energy and moisture are transferred to the medium. As the medium rotates into the opposing air stream, the warmed, moist medium transfers the heat and moisture to the opposite-flowing air stream. Therefore, a heat wheel can either: reduce entry of warm, moist outside air into the lab building, or recover heat and moisture that would have been simply exhausted for the building. There has been a renewed interest in heat wheels since molecular sieve coatings have been used that ensure minimal contaminant transfer

Direct Connected Centrifugal Fan Design

Application

Y5-11 Series Boiler centrifugal fan are applicable to industrial boilers that burn a variety of media and equipped with devices to reduce smoke and dust. The temperature of the transmitted medium can not extend to 250°c, and the total pressure power can reach to 85%.

Types

Y5-11 Series Centrifugal fans can be devided into13 specifications. They are No.3 to No.24.

Structure

The blowers mainly consist of impeller, case, inlet port, air control unit and transmission part

Function Table Information for Ordering

When you make an order for our products,please note clearly about the type,volume, pressure, outlet angle,rotational direction,and the specifications of the motor of fan.

Applications

Type High Pressure Centrifugal Fan Design For Cement Kiln Use is generally used in Cement kiln. The medium carried by centrifugal fan must below 80°C .

The solid particle and duct in the medium is less than $150m/m^3$.

Types and Explanation

High Pressure Centrifugal Fan are all single suction, the driving mode is A Type drive. The outlet of the fan is shown by outside of case that are all be made into 0°, 45°, 90°, 135°, 180° five kinds of angle.

Structure

The type of fans mainly consist of impeller, case, inlet port, pedestal and so on.

1. Impeller: It is designed by high efficiency forward fan theory and the molding impeller can be stable operation after static and dynamic balancing
2. Main case: Volute case is welded by general steel plate.
3. Inlet: It is welded by general steel plate to convergence and steamlinc structure. Good liquidity.

Housing for a Centrifugal Fan

A housing for a blower, fan or pump or turbine, the housing adapted to be associated with a rotor adapted in use to cooperate with fluid flowing through the housing wherein the housing comprises a shroud for guiding the fluid moving in association with the rotor, the rotor having at least one vane adapted to cooperate with the fluid to drive or to be driven by the fluid, wherein the shroud is configured to promote vortical flow of the fluid through the housing.

Claims

A housing for a rotor, the housing comprising: an inlet portion comprising a shroud for guiding a fluid moving in association with the rotor; an outlet portion to exhaust the fluid, the outlet portion extending the shroud from the inlet portion, the shroud expanding axially away from a region of rotation of the rotor, the shroud promoting vortical flow of the fluid through the housing.

The housing of claim 1, wherein the configuration of the shroud corresponds to the streamlines of a vortex.

The housing of claim 1, wherein the configuration of the shroud conforms to a two-dimensional spiral.

The housing of claim 1, wherein the configuration of the shroud conforms to a three-dimensional spiral.

The housing of claim 1, wherein the shroud includes an active surface configured to guide the fluid flowing within the housing, the active surface having a configuration conforming substantially to that of a logarithmic spiral.

The housing of claim 5, wherein the logarithmic spiral unfolds at a constant order of growth when measured at equiangular radii.

The housing of claim 1, wherein the shroud includes an active surface configured to cooperate with the fluid flowing within the housing, the active surface having a configuration conforming substantially to that of a logarithmic curve.

The housing of claim 7, wherein the logarithmic curve unfolds at a constant order of growth when measured at equiangular radii.

The housing of claim 1, wherein the outlet portion extends a flow path for a fluid in a continuous manner from the inlet portion.

The housing of claim 1, wherein the shroud substantially surrounds at least the perimetre of the rotor and provides a space between an inner surface of the shroud and a surface swept by an outer edge of at least one vane of the rotor during rotation of the rotor.

The housing of claim 10, wherein the space increases from a minimum cross-sectional area to an expanded cross-sectional area.

The housing of claim 1, wherein the inlet portion comprises a mount to support the rotor.

The housing of claim 1, wherein the flow of the fluid is redirected in a generally transverse direction relative to the incoming flow.

The housing of claim 1, wherein the shroud comprises a separate component from an external surface of the housing.

The housing of claim 1, wherein the housing is adapted to a profile of the rotor.

The housing of claim 14, wherein the profile of the rotor is that of an axial rotor.

The housing of claim 14, wherein the profile of the rotor is that of a centrifugal rotor.

The housing of claim 14, wherein the profile of the rotor is intermediate to that of an axial rotor and that of a centrifugal rotor.

A rotor housing apparatus for use with a centrifugal rotor, comprising: a shroud enclosing a region of rotation for a rotor; an inlet located on a first portion of the shroud; an outlet located on a second portion of the shroud, the second portion of the shroud being approximately opposed with respect to the first portion of the shroud, wherein an internal surface of the shroud includes a vortical formation that induces a vortical flow in a fluid traversing a fluid pathway between the inlet and the outlet.

A rotor housing apparatus for use with an axial rotor, comprising: a shroud enclosing a region of rotation for a rotor; an inlet located on a first portion of the shroud; an outlet located on a second portion of the shroud, the second portion of the shroud being approximately opposed with respect to the first portion of the shroud, wherein an internal surface of the shroud includes a vortical formation that induces a vortical flow in a fluid traversing a fluid pathway between the inlet and the outlet.

Description

Cross-reference to Related Applications

The present application is a continuation and claims the priority benefit of U.S. patent application Ser. No. 11/496,013 filed Jul. 28, 2006, which is a continuation and claims the priority benefit of Patent Cooperation Treaty application number PCT/AU2005/000116 filed Jan. 31, 2005, which claims the priority benefit of U.S. provisional patent application Nos. 60/540,513 filed Jan. 30, 2004; 60/608,597 filed Sep. 11, 2004; and 60/624,669 filed Nov. 2, 2004. The disclosure of the aforementioned applications is incorporated herein by reference.

Background

Field of the Invention

The present invention relates to a housing or chamber for a fan for moving air, pump for inducing fluid flow or torque generator, which is responsive to fluid flow such as a turbine. In particular it is directed to providing an improved housing for such apparatus to improve the efficiency of such devices.

Description of the Related Art

Centrifugal fans, blowers, pumps turbines and the like represent approximately half of the world's fan, pump and turbine production each year. As fans or pumps, they are used to produce higher pressure and less flow than axial impellers and fans. They are used extensively where these parameters must be satisfied. They have also been used advantageously where installation limitations might not permit an axial fan to be used. For example, applications such as domestic exhaust fans require greater flow with a relatively low pressure difference. Such an application would normally be satisfied by an axial type of fan. However, in many cases, a centrifugal fan is used to turn the flow path at right angles so that it can fit into a roof or wall cavity. An axial fan will not fit into the cavity and maintain efficiency. In another example, the exhaust ducting in many buildings is only 3 or 4 inches in diametre. It is impractical to fit an effective high-output axial fan to such a small duct.

While centrifugal fans have been used for a long time, little attention has been given to the design of the housing in which the rotor is retained. Where issues of efficiency and noise are investigated, the designer's attention is given primarily to the impeller. Historically, such housings have not been optimised for: 1. fluid flow drag reduction; 2. noise reduction; 3. adjustment of the pressure/flow relationship. Additionally, the housings of typical centrifugal fans, blowers, pumps turbines and the like cause the incoming fluid to turn sharply before leaving the housing. Such shapes are detrimental to efficient performance of the device overall, often introducing significant turbulence.

In the previous disclosure of the applicant for a Fluid Flow Controller as published in W003056228, the applicant has noted the benefits that can be obtained by allowing fluid to flow in the manner followed in Nature.

Summary of the Invention

Exemplary housings for a rotor are provided. In some embodiments, a housing includes an inlet portion with a shroud for guiding a fluid moving in association with the rotor, and an outlet portion to exhaust the fluid, the outlet portion extending the shroud from the inlet portion, the shroud expanding axially away from a region of rotation of the rotor, the shroud promoting vortical flow of the fluid through the housing. Such housings may be adapted for an axial rotor, a centrifugal rotor, or a rotor having a profile intermediate to that of an axial rotor and that of a centrifugal rotor.

In some embodiments, a rotor housing may include a shroud enclosing a region of rotation for a rotor, an inlet located on a first portion of the shroud and an outlet located on a second portion of the shroud, the second portion of the shroud being approximately opposed with respect to the first portion of the shroud, wherein an internal surface of the shroud includes a vortical formation that induces a vortical flow in a fluid traversing a fluid pathway between the inlet and the outlet. Such housings may be adapted for an axial rotor, a centrifugal rotor, or a rotor having a profile intermediate to that of an axial rotor and that of a centrifugal rotor.

Detailed Description

Each of the embodiments is directed to a housing for a fan, blower, pump or turbine or the like, which provides an efficient fluid pathway. Hereinafter in this description the term 'fan' will be used generically to refer to any fan, blower, pump, turbine or the like. Where a reference is made to a fan driving or promoting fluid flow, it is to be appreciated that the reference is intended to encompass the situation where the fluid flow drives a rotor of a turbine or the like.

In order to appreciate the differences from the prior art, it is helpful to describe the key features of housings conventionally used for centrifugal fans. An example is illustrated diagrammatically, which illustrates the key features of a typical arrangement of a housing 1 for a centrifugal fan. Conventionally, such a housing 1 is configured in shape to follow the form of a spiralling arc in two dimensions. It generally comprises a pair of flat-sided panels 3 and 4 disposed apart, parallel to each other and sealed around the perimetre by an edge panel 5 formed from a planar sheet. This creates angled comers 6 at the junction between the top panel 3 and edge panel 5 and similarly between the bottom panel 4 and edge panel 5. Such angled comers induce unwanted turbulence in the fluid passing within the housing.

The shape of a spiralling arc means that a space is provided between the inner surface of the edge panel and the imaginary surface swept by the outer edges of the vanes of the rotor. It will be appreciated that the depth of this space increases progressively from a minimum to a maximum through an angle of 360 degrees. In the vicinity of the maximum depth an outlet is provided to exhaust the fluid.

Each of the embodiments is directed to a housing for a fan, which provides an efficient fluid pathway for fluid passing through the housing. Such fans comprise a rotor which is normally provided with a plurality of vanes or blades although a rotor having a single blade is possible. The vanes are generally configured to provide an outward or radial component of acceleration to the fluid being driven, or in the case of a turbine, the fluid is deflected to provide a radial component to the force applied to the vane and thereby a deflection to the fluid including a radial component.

Nature provides excellent models of optimised streamlining, drag reduction, and noise reduction. Any biological surface grown or eroded to optimise streamlining has no angled corners and does not make fluid turn at right angles but generally follows the shape of an eddy constructed in accordance with a three-dimensional equiangular or Golden Ratio spiral. The underlying geometry of this spiral is also found in the design of a bird's egg, a snail, and a sea shell.

These spirals or vortices generally comply with a mathematical progression known as the Golden Ratio or a Fibonacci like Progression.

Each of the embodiments, in the greater part, serves to enable fluids to move in their naturally preferred way, thereby reducing inefficiencies created through turbulence and friction which are normally found in housings for centrifugal fans.

Previously developed technologies have generally been less compliant with natural fluid flow tendencies.

It has been found that it is a characteristic of fluid flow that, when it is caused to flow in a vortical motion through a pathway that the fluid flow is substantially non-turbulent and as a result has a decreased tendency to separate or cavitate. It is a general characteristic of the embodiments that the housings described are directed to promote vortical flow in the fluid passing through the housing. It has also been found that vortical flow is encouraged where the configuration of the housing conforms to a two-dimensional or three-dimensional spiral. It has further been found that such a configuration tends to be optimised where the curvature of that spiral conforms substantially or in greater part to that of the Golden Section or Ratio. It is a characteristic of each of the embodiments that the greater proportion of the internal surfaces which form the housing have a curvature which takes a two dimensional or three dimensional shape approaching the lines of vorticity or streak lines found in a naturally occurring vortex. The general form of such a shape is a logarithmic spiral. It has further been found that the performance of the embodiments will be optimised where the curvature of the surfaces of the housing substantially or in the greater part conform to the characteristics of the Golden Section or Ratio. It has further been found that the performance is optimised if any variation in cross-sectional area of the fluid pathway also substantially or in greater part conforms to the characteristics of the Golden Section or Ratio.

It has also been found fluid flow is more efficient if the surfaces over which the fluid flows have a curvature substantially or in greater part correspond to that of the Golden Section. As a result of the reduced degree of turbulence which is induced in the fluid in its passageway through such a fan, the housing according to the various embodiments can be used for conducting fluid with less noise and wear and with a greater efficiency than has previously been possible with conventional housing of equivalent dimensional characteristics.

The greater percentage of the internal surfaces of the housings of each of the embodiments described herein are generally designed in accordance with the Golden Section or Ratio and therefore it is a characteristic of each of the embodiments that the housings provides a fluid pathway which is of a spiralling configuration and which conforms at least in greater part to the characteristics of the equiangular or Golden Section or Ratio. The characteristics of the Golden Section are illustrat which illustrates the unfolding of the spiral curve according to the Golden Section or Ratio. As the spiral unfolds the order of growth of the radius of the curve which is measured at equiangular radii (e.g. E, F, G, H, I and J) is constant. This can be illustrated from the triangular representation of each radius between each sequence which corresponds to the formula of a:b=b:a+b which conforms to the ratio of 1:0.618 approximately and which is consistent through out the curve.

This invention may, alternatively, use a snail or sea shell-like shaped flow path housing which may be logarithmic but not a Golden Ratio. Although it is not optimised if it doesn't conform to the three-dimensional Golden Ratio, it will still provide superior performance in its intended use over conventional designs.

The fan assembly 11 comprises a fan rotor 12 having a plurality of vanes 13, the rotor 12 being adapted to be rotated by an electric motor, not shown. The fan motor is supported within a housing 14 having an inlet 16 and an outlet 17.

The housing 14 has a whirl-shaped form, at least on the internal surfaces which resembles the shape of shellfish of the genus Trochus. This shape corresponds generally to the streamlines of a vortex. In the drawings it is to be appreciated that the form indicated on the external surfaces is intended to correspond with the form of the internal surface, although in a real fan the form of the external surface is not of importance to the performance of the fan as such and may be quite different from the internal surfaces. Indeed, the housing might be constructed with an internal shroud which comprises a separate component from the external surface of the housing, and it is to be appreciated that where such a design is undertaken, it is the internal surfaces of the separate shroud which must conform to the principles as described herein.

In the first embodiment, the housing is formed in two portions, 18 and 19. The first of these comprises an inlet portion 18 which includes the inlet 16 and also provides mounting means (not shown) to support the fan motor to which the fan rotor 12 is attached. The inlet portion 18 also acts as a shroud around outer extents of the vanes 13 of the rotor 12 and provides a space 22 between the inner surface 21 of the inlet portion 18 and the imaginary surface swept by the outer edges 23 of the vanes 13 during rotation of the rotor 12. It will be seen that the depth of this space increases between a minimum space 25 and a maximum space in a manner akin to the corresponding space in a conventional centrifugal fan. Unlike a conventional centrifugal fan, however, this increase in the space is accompanied by displacement of the fluid path axially away from the region of rotation of the rotor in the first portion 18 towards the outlet 17. The second portion of the housing 14 comprises an outlet shroud 19 extending the flow path in a continuous manner from the first portion. In the outlet shroud 19, the inner surface of the shroud 19 continues to expand while the fluid path is displaced axially. As a result, a generally vortically shaped fluid path is provided which urges fluid flowing through the housing 14 to adopt a vortical flow pattern, as indicated by the dotted line 27. Such a flow pattern is of higher efficiency and lower noise than for a comparable conventional fan. In addition, by being spun into vortical flow, the fluid may be urged to be redirected in a generally transverse direction relative, to the incoming flow without requiring an abrupt and turbulent change in flow direction. This also improves efficiency and reduces noise.

As mentioned earlier, while a housing having a generally vortical internal form can be expected to provide significant improvements in higher efficiency and reduced noise, the benefits will be optimised by configuring the housing to have a vortical form in the nature of a three dimensional equi-angular spiral or "Golden- Section" spiral. Such a shape should have the internal surfaces configured to have a curvature conforming to the Golden Section. Such a shape will conform with the natural flow tendencies of fluids, thereby further improving efficiency.

It is to be appreciated that the configuring of the housing to be in two portions is to provide ease of manufacture, assembly and maintenance, only. The two portions of such a housing may be held together by releasable clasping means such as clips (not shown), or may include cooperating flanges, bayonet fastenings, or other suitable joining means.

The first embodiment is adapted so that the housing 31 may be manufactured as a single piece, for example, by rota-moulding. Alternatively, the housing may comprise more than two portions.

A third embodiment of a fan 41 comprising a rotor 42 having a single vane 43 having an expanding screw-like form. This rotor 42 is accommodated within an extended housing 44, which nevertheless takes a vortical form. It is envisioned that such a design may be appropriate for more viscous fluids or fluid-like materials.

While a housing according to the first and second embodiments will provide improved performance when used with rotors having a wide range of vane configurations, it is to be appreciated that performance of the fan assembly will also depend on the configuration of the rotor. It has been found that performance may be further improved where the rotor itself is designed to provide flow in accordance with the principles of nature. Such a rotor is described in the applicants co-pending application entitled "Vortical Flow Rotor." It is to be understood that such a rotor is directed to providing a vortical flow stream, and when appropriately configured in conjunction with a housing according to the first or second embodiment, an optimised performance characteristic can be achieved.

It can be understood in light of the above description that a housing according to the first and second embodiments will provide performance improvements where a centrifugal rotor is used. It can be seen that the application of a radial component of fluid flow to the flow stream, will urge fluid outwardly as well as rotationally, thereby adopting a vortical flow. It is not so obvious that use of a housing of the first embodiment with an axial fan will also provide a significant performance improvement, yet this has been found to be the case. It seems that the provision of a housing that easily accommodates vortical flow promotes such vortical flow in practice. Therefore, it is within the cope of the invention now disclosed that the housing may be used with a rotor axial configuration.

This discovery has led to a further advance. The vanes of the rotor that can be used within the housing of the first embodiment may be configured with a profile that is intermediate between an axial and a centrifugal rotor. As mentioned earlier, axial and centrifugal rotors have quite differing performance characteristics: the axial rotor promoting high flow at low pressure while the centrifugal rotor promotes low flow at high pressure. By selecting a rotor with an intermediate characteristic, the performance of the fan can be "tailored" to more precisely match the application. The precise configuration of the housing may also be "tuned" to cooperate fully with the selected rotor to even further improve the design characteristics. Such flexibility has not been appreciated previously.

A designer can now approach a project knowing that he can properly design an appropriate fan for the task, rather than adopting an inappropriate fan due to physical constraints.

Additionally, it has been found that the compound curves of the housing of the above embodiments have rigidity and structural integrity considerably beyond flat sided panels found in conventional housings and thereby can be built from lighter and thinner materials. Nevertheless, the inherent stiffness, combined with the lack of turbulence within the fluid flow also reduce noise—a major problem in conventional housings. Flat-sided housings vibrate, drum, resonate, and amplify noise. The housing of the embodiments reduces vibration, drumming, resonance, and amplification of noise. While it is believed that a fan having superior performance will generally be achieved by designing the housing in a three-dimensional vortical form as described in relation to the first embodiment, there will be instances where it will not be practicable to adopt such a form. This is more likely to be the case where the fan is to be used in an existing installation that has previously incorporated a conventional centrifugal fan. Nevertheless, significant improvements can be obtained by incorporating into the design of a conventional centrifugal fan the principles revealed in the first embodiment.

When assembled together, the first and second halves provide a fluid space between the internal surface of the housing and the imaginary surface swept by the outer edges of the vanes 13 during rotation of the impeller 69. This space increase from a minimum at a point "A" to a maximum at an adjacent point "B."

At the maximum point "B" the housing incorporates an outlet opening 71 transverse to the plane of rotation of the impeller which is co-planar with the axis. In use an outlet duct 72 (as shown in dotted lines) will normally be mounted to the outlet to convey the fluid from the housing.

The walls of the two halves around the space are curved with a curvature which substantially conforms with the Golden Section. This curvature is also be configured to cause the fluid to flow within the space in a spiralling, vortical motion. As a result, drag in the fluid flow through the space is reduced.

Throughout the specification, unless the context requires otherwise, the word "comprise" or variations such as "comprises" or "comprising," will be understood to imply the inclusion of a stated integer or group of integers but not the exclusion of any other integer or group of integers.

Centrifugal Fans

OVERVIEW

Industrial *Centrifugal Fans* differ from *Industrial Axial Flow Fans* in the way they move the air. A*Centrifugal Fan (or Centrifugal Blower)* typically has one of seven types of *wheels* that draws the air into the inlet of the blower housing, through the wheel, and discharges it at 90 degrees out through the discharge of the housing. An *Axial Flow Fan* or *Industrial Axial Fan* uses a *propeller* to draw the air into it and discharges it in the same axial direction. An "air moving device" with a propeller is typically referred to as a *Fan*, while one with a wheel is typically referred to as a *Blower*.

Making the proper centrifugal fan selection, involves considering the following conditions the blower will be subjected to:

1. *What is the airflow that will be required?*

Airflow is rated in cubic feet of air per minute (CFM) or in metric equivalent, it is rated in cubic metres per hour (M^3/Hr).

2. *What is the static pressure through the complete system?*

This is rated in static pressure water gauge (SPWG) or in metric equivalent, it is rated in Pascal's (Pa). SPWG or Pa is the resistance to flow, or friction, caused by the air moving through a pipe or duct. Be sure to include losses through filters, dampers, etc.

NOTE: If the static pressure on the inlet side of the blower will exceed 15" SPWG, a rarefication and a suction pressure correction chart.

3. *What is the ambient air temperature outside the blower?*
4. *What will be the airflow conditions through the blower?*
    a. Clean or dirty (contaminated) air?
    b. Dry or wet (moisture laden) air?
    c. Temperature of air?
5. *What is the altitude the blower will be operating at?*
6. *Is anything going through the blower that could be explosive or flammable?*
7. *Will the general location the blower will be operating in contain any type of explosive or flammable material or liquid?*
8. *If the blower will be used to convey material, answer the following:*
    a. What is the material?
    b. How much does the material weigh, in pounds per cubic foot?
    c. How many pounds will you need to convey per hour?
    d. Is the material long and stringy or granular?

Most Centrifugal Fans (Blowers) can be used for any of the above parameters except, if conveying material, a blower with a forward curve (squirrel cage) wheel is not suitable. The best centrifugal wheel type for conveying material is a radial wheel or an open radial wheel. If the material you will be conveying is log and stringy, like paper trim or fibres, only an open radial wheel should be used. The best models to use, for conveying material through the fan, are Models PB, SPB or PBS Pressure Blowers and Model RBE Radial Blade Exhauster. Models HDBI (Heavy Duty Backward Inclined Blowers) and HDAF (Heavy Duty Airfoil Blowers) are also used in many dust collector applications where the blower is on the clean side of the dust collector.

If the blower will be used to create suction, pressurisation, cooling, combustion air, forced ventilation or general ventilation, then only the first seven conditions above need to be considered. After the first seven conditions are known, the blower selection will be based on the type of blower that is best suited by performance, design and features that best meet those conditions.

Cincinnati Fan does not manufacture any consumer or residential type centrifugal fans. But, we do manufacture 14 models of Centrifugal Fans (Blowers) that are designed and constructed as*industrial* type. We manufacture centrifugal blowers in the ¼ to 125 horsepower range, with airflow capacities from 50 to 51,900 CFM and static pressures from 1/8" to 78" SPWG.

Applications

While a list of all the applications our Centrifugal Fans are used for would be extremely long, the most common include:

- Conveying material in dust collector systems or loading to silos, trucks and railroad cars.
- Combustion air for burners and industrial water heaters.
- Cooling electrical equipment or parts from heat curing ovens and kilns.
- Pressurising cabinets, vaults or rooms.
- Blow-off systems to remove moisture from parts prior to painting, coating or packing.
- General ventilation of rooms or factories.
- Providing suction or pressure for industrial process air flotation tables.
- Drying ink on cans and bottles, silk screen and printing processes.

- Food processing.
- Circulating air in ovens and dryers to eliminate cold spots.
- Forced ventilation of AC and DC motors, traction motors and electrical cabinets.

Benefits

The benefits of Cincinnati Fan Centrifugal Fans are:

- Most all types and models, *with most accessories*, are shipped within 10-15 working days. Some models available on a 5 Day RUSH program, *with most accessories. Quicker than industry standards.*
- Spark resistant, cast aluminium wheels are *standard* in our PB and SPB Models. *Ideal for applications requiring spark resistant construction with quick deliveries.*
- All fabricated steel models have continuously welded housings for additional strength. *Superior strength over stitch welding or lock seams.*
- Heavy duty, industrial components and continuous duty motors ensures long lasting operation minimising equipment down time. *Dependable service with few field problems.*
- All aluminium or all stainless steel fan construction is "not special" for us. We manufacture most models with both types of construction almost weekly. *Faster delivery to you.*
- All of our Industrial Blowers are available with multiple wheel sizes and some are also available with multiple housing widths from 50% to 100%, in 5% increments. Direct driven fans can be selected instead of larger, heavier, more expensive belt driven fans. *Less expensive, more compact and lighter weight. Design flexibility.*
- Belt driven models utilise wider spacing between the fan bearings which increases the bearing life. *Less down time before replacement bearings are needed.*
- All fans and blowers are manufactured in our 110, 000 square foot, state-of-the-art factory in Mason, Ohio. *Not imported.*
- *Service!* Without it, all the above is meaningless. We pride ourselves and are highly respected in the fan industry for the quality and service we give all our customers. *Let us prove it to you also.*

The design steps, listed previously, are critical for the proper selection and construction of your centrifugal fan. We have over 170 experienced sales engineers across the U.S.A. and Canada ready to assist you in making the proper fan or blower selection. Call our sales office for your area. They most likely have already sold some blowers for your type of application.

Axial Flow Fan, or Industrial Axial Fan

Overview

Axial Flow Fans are different from *Centrifugal Fans and Blowers* in the way they move the air. An *Axial Flow Fan* or *Industrial Axial Fan* uses a propeller to draw the air into it, and discharge it, in the same axial direction. A *Centrifugal Blower* draws the air into the inlet of the blower housing with one of seven types of *wheels*. The blower draws the air through the wheel, and discharges it at 90 degrees out through the discharge of the housing. A device that moves air with a propeller is typically referred to as a *Fan*, while one with a wheel is referred to as a *Blower*.

In this section we will cover axial type *Industrial Axial Fans* or *Industrial Axial Flow Fans*.

The two most common types of industrial, *Axial Flow Fans* are *Tube Axial Fans* (or Ducted Fans) and Panel *Fans* or *Wall Fans. Tube Axial Fans* use a propeller, rotating in a round section of duct, which is mounted into ductwork of the same diametre to exhaust hot and/or contaminated air from a process or system. Panel Fans or *Wall Fans* use a propeller, rotating in a flat panel, which is mounted to a wall to exhaust air from a room or draw air into a room. Other *Axial Flow Fans* that incorporate the Tube *Axial Fan* design are *Personnel Coolers and Air Circulators.*

Tube Axial Fans can be direct driven with the propeller mounted directly on the motor shaft or belt driven with the propeller mounted on a separate fan shaft and driven by the motor through belts and sheaves. Axial Flow Fans are ideal for moving large volumes of air, rated in cubic feet of air per minute (CFM) at low pressures, rated in static pressure (SP).

The propellers used in Axial Flow Fans are typically made of plastic, fabricated steel, cast aluminium or stainless steel. However, plastic propellers are generally not considered “industrial duty” and are therefore used only in commercial or residential fans.

To make the proper fan selection, you must consider the following system conditions the fan will be subjected to:

1. *What is the required airflow?*

   This is rated in cubic feet of air per minute (CFM) or, in metric equivalent, it is rated in cubic metres of air per hour (M^3/Hr).

2. *What is the static pressure through the complete system?*

   This is rated in static pressure water gauge (SP) or, in metric equivalent, it is rated in Pascal's (Pa).

3. *What is the ambient air temperature outside the fan?*

4. *What will be the airflow conditions through the fan?*

   a. Clean or dirty (contaminated) air?

   b. Dry or wet (moisture laden) air?

   c. Temperature of air?

5. *What is the altitude the fan will be operating in?*

6. *Will the air stream, inside the fan, contain anything that could be explosive or flammable?*

7. *Will the general location the fan will be operating in contain any type of explosive or flammable material or liquid?*

Cincinnati Fan offers five types of industrial Axial Flow Fans. All include heavy-duty designs for industrial applications. We manufacture Tube Axial Duct Fans, Panel Fans, Personnel Coolers and Air Circulators. Our Tube Axial Duct Fans are available in both direct driven or belt driven models. Our Panel Fans, Personnel Coolers and Air Circulators are all direct driven. All models are available with cast aluminium, steel or stainless steel propellers. All models can also be manufactured in all aluminium or all stainless steel construction. Cincinnati Fan does not manufacture any consumer or residential type fans. But, we do offer 18 models of *Axial Flow Industrial Fans.* We manufacture fans in the ¼ to 15 horsepower range with airflow capacities from 700 to 44,500 CFM and static pressures from 1/8" to 1-1/8" SPWG.

Applications

The most common applications for *Axial Flow Fans* include:

- Paint spray booth exhaust.
- Fume hood exhaust to exhaust dangerous fumes away from the operator.

- Welding fume exhaust in welding booths.
- Process cooling and exhaust of machinery and/or systems.
- Spot cooling of heat generating equipment such as generators or transformers.
- General ventilation using supply air and/or make-up air in factories, warehouses, foundries, garages, laundries and transformer, engine or equipment rooms.
- People cooling in hot work areas.
- Exhausting heat and cooling of kilns, forges and in steel mills.
- Boosting airflow through long ductwork.
- Exhausting mist and vapour in paper mills or industrial parts washers prior to painting parts.

Benefits

The benefits of Cincinnati Fan Axial Flow Fans are:

- Most types and models, with most accessories, are shipped within 10-15 working days. *Quicker than industry standards.*
- Spark resistant, cast aluminium propellers are standard. Not optional or extra as with some other fan manufacturers. *Ideal for paint spray booth exhaust.*
- All aluminium or all stainless steel construction is "not special" for us. We manufacture most models with both types of construction almost weekly. *Faster delivery to you.*
- All types, models and sizes are available with multiple propellers for a wider range of air flows to meet your exact requirements. *More design flexibility.*
- Heavy duty, industrial components and continuous duty motors ensures long lasting operation minimising equipment down time. *Dependable service with few field problems.*
- Belt driven models utilise wider spacing between the fan bearings which increases the bearing life before replacements are needed. *Less down time.*
- All fans and blowers are manufactured in our 110,000 square foot, state-of-the-art factory in Mason, Ohio. *Not imported.*
- *Service!* Without it, all the above is meaningless. We pride ourselves and are highly respected in the fan industry for the quality and service we give all our customers.

Let us prove it to you also. The design steps, listed previously, are critical for the proper selection and construction of your fan. We have over 170 experienced sales engineers across the U.S.A. and Canada ready to assist you in making the proper fan selection. Call our sales office for your area. They most likely have already sold some fans for your type of application

Industrial Fans & Blowers

Overview

Centrifugal fans or blowers have one of seven types of wheels, normally mounted in a fan or blower housing. The air enters the housing inlet, is turned 90 degrees and is exhausted out of the housing discharge. An Axial fan normally uses a propeller, having 2 or more blades, to draw the air through a section of ductwork or a flat wall panel and out the other side in an axial direction. There is also one type of Axial fan that has a wheel mounted in a section of duct work and is classified as a Tubular Centrifugal.

We will use the term "Fan" for either a fan or blower. To make the proper fan selection, you must consider the following conditions the fan will be subjected to:

1. What is the airflow that will be required. This is rated in cubic feet of air per minute (CFM) or in metric equivalent, it is rated in cubic metres per hour (M^3/Hr).
2. What is the static pressure through the complete system. This is rated in static pressure water gauge (SPWG) or in metric equivalent, it is rated in Pascal's (Pa).

   Note: SPWG or Pa is the resistance to flow, or friction, caused by the air moving through a pipe or duct. Be sure to include losses through filters, dampers, etc.
3. What is the temperature of the air going through the fan?
4. What is the ambient air temperature outside the fan?
5. Will the air stream be clean, dry air?
6. If the fan will be used to convey material, answer the following:

a. What is the material?

b. How much does the material weigh, in pounds per cubic foot?

   c. How many pounds will you need to convey per hour?
   d. Is the material long and stringy or granular?
7. Will there be any moisture in the air stream?
8. What is the altitude the fan will be operating at?
9. Is anything going through the fan that could be explosive?
10. Does the location the fan will be operating in contain any explosive material, liquid or both?

An Axial fan can be used for all the above parameters, except #6 above, if you will be using the fan to convey material. Most Centrifugal fans can be used for any of the above parameters except if conveying material a fan with a forward curve (squirrel cage) wheel is not suitable. The best centrifugal wheel type for conveying material is a radial wheel or an open radial wheel. If the material you will be conveying is log and stringy, like paper trim or fibres, only an open radial wheel should be used.

Cincinnati Fan does not manufacture any consumer or residential type fans or blowers. But, we do manufacture 35 models of axial fans and centrifugal blowers that are designed and constructed as *industrial* type. We manufacture fans and blowers in the ¼ to 125 horsepower range, with airflow capacities from 50 to 48,500 CFM and static pressures from 1/8" to 78" SPWG.

Applications

While a list of all the applications industrial fans and blowers are used for would be extremely long, the most common are:

Centrifugal Fans & Blowers

- Conveying material in dust collector systems or loading to silos, trucks and railroad cars.
- Combustion air for burners and industrial water heaters.
- Cooling electrical equipment or parts from heat curing ovens and kilns.
- Pressurising cabinets, vaults or rooms.
- Blow-off systems to remove moisture from parts prior to painting, coating or packing.
- General ventilation of rooms or factories.
- Providing suction or pressure for industrial process air flotation tables.

- Drying ink on cans and bottles, silk screen and printing processes.
- Food processing.
- Circulating air in ovens and dryers to eliminate cold spots.
- Forced ventilation of AC and DC motors, traction motors and electrical cabinets.

Industrial Axial Fans

- Paint spray booth exhaust.
- Fume hood exhaust.
- Exhausting dangerous fume away from the operator.
- Welding fume exhaust for welding booths.
- Process cooling and exhaust of machinery and/or systems.
- People cooling in hot work areas.
- Forced cooling and exhausting heat in steel mills.
- Forced cooling for kilns, ovens and forges.
- Exhausting mist and vapour in paper mills or industrial parts washers prior to

painting parts.

- General ventilation using supply air and make-up air in factories, foundries and warehouses.

Call us or our sales office for your area. We most likely have already built industrial fans and blowers for your application.

Benefits

The benefits of Cincinnati Fan industrial fans and blowers are:

- Most all types and models, *with most accessories*, are shipped within 10-15 working days. Some models available on a 5 Day RUSH program, *with most accessories. Quicker than industry standards.*
- Spark resistant, cast aluminium wheels are *standard* in our PB and SPB Models and cast aluminium props are *standard* in most of our Axial Fan Models. *Ideal for applications requiring any spark resistant requirement and quick deliveries.*
- Heavy duty, industrial components and continuous duty motors ensures long lasting operation minimising equipment down time. *Dependable service with few field problems.*

- All aluminium or all stainless steel fan construction is "not special" for us. We manufacture most models with both types of construction almost weekly. *Faster delivery to you.*
- All of our centrifugal blowers and fans are available with multiple wheel sizes and some are also available with multiple housing widths from 50 to 100% in 5% increments. Direct driven fans can be selected instead of larger, heavier, more expensive belt driven fans.

Less expensive, more compact and lighter weight. Design flexibility.

- Belt driven models utilise wider spacing between the fan bearings which increases the bearing life. *Less down time before replacement bearings are needed.*
- *Service!* Without it, all the above is meaningless. We pride ourselves and are highly respected in the fan industry for the quality and service we give all our customers. *Let us prove it to you also.*

Chapter 8

Exhaust Blowers

A typical axial type Ventilation Fan will work okay for applications that don't require long hose or duct runs. But, if the exhausted air has to be moved a long distance through hose or duct work, an Exhaust Blower will be needed to supply the proper exhaust air flow at the higher static pressures required.

An exhaust blower is a centrifugal type blower having a housing that contains a blower wheel of some type. The relationship between the housing scroll and the wheel configuration allows a blower to produce much higher pressures than a typical axial ventilation fan.

For the different types of Exhaust Blowers manufactured by Cincinnati Fan, their designs and specifications go to Centrifugal Blowers & Fans.

Portable Fume Exhausters: Portable Blowers

Cincinnati Fan manufacturers eight sizes of "Portable Fume Exhauster-Blowers". Five "EBR" models are standard with spark resistant, self-cleaning, cast aluminium, radial wheels and steel blower housings. Three "EBM" models have multi-vane, forward curve wheels and cast aluminium blower housings. "EBR" models are AMCA Type B spark resistant. "EBM" models are AMCA Type C spark resistant. All models come standard with continuous duty (not duty rated), TEFC motors. Explosion Proof (EXP) motors are available for most models. The spark resistant feature of our Fume Exhauster-Blowers, with EXP motors, is a requirement for explosive environments. All units are self-contained with 10 feet of hose and intake nozzle. A unique feature to "EBR" and "EBM" Exhauster-Blowers is that the hose will fit on either the inlet or outlet of each model to exhaust fumes or blow fresh air. They can be mounted stationary to a table, wall or beam or they can be carried (or mounted on a dolly) as portable units for where ever needed. The ¼ to 1 horsepower, 1 Phase, 60 Cycle, models (with TEFC motors) also include an 8 foot cord, 3 prong grounded plug and switch pre-wired for 115 Volt.

Applications

A few of the many applications for this product include:

- To exhaust injurious welding, brazing, soldering, etching or engraving fumes away from the operator and outside the building.
- Supplying fresh air to workers inside confined areas such as tanks, vats, or man holes.

Features

- AMCA Type B or C, spark resistant construction standard for all models.
- Heavy-duty 14 gauge steel blower housing on all "EBR" models. Housings are black powder coated for long life. "EBM" models have cast aluminium blower housings.
- Rugged 14 gauge steel base with black powder coating resists rusting.
- Tote handle with soft foam grip for ease of moving.
- Three rubber pads under base to eliminate creeping and vibration during operation.
- Totally Enclosed Fan Cooled (TEFC) motors are standard.
- Motors are continuous duty, full horsepower rated. NOT duty rated.
- All models available with 115/230 Volt, 1 Phase, 60 Cycle or 230/460 Volt, 3 Phase, 60 Cycle motors. 575 Volt and 50 Cycle motors also available.
- Exp motors available for most models.
- Single Phase, TEFC motors (up to 1 HP) come with an 8 foot cord, 3 prong grounded plug and a switch pre-wired for 115 Volt, 60 Cycle.
- Suction nozzle and ten feet of hose included as standard.
- One size hose fits inlet or discharge.
- Inlet and discharge guards, required by OSHA, are standard
- Safe, no exposed moving parts.

Benefits

Benefits of Cincinnati Fan "Fume Exhauster-Blowers" are:

Location

Fume Exhauster frame can be bolted to a table, wall or beam as a permanent location. Or they can be carried, or bolted to a two wheeled dolly or dolly cart for portability from one location to another. *More flexibility.*

Compact & Fewer Parts

Units are very compact and rugged. One size hose fits the inlet OR discharge of each blower housing. *Built to last and no need for multiple hose sizes.*

Spark Resistant Construction

"EBR" models are AMCA Type B spark resistant and "EBM" models are AMCA Type C spark resistant, so they can be used in explosive atmosphere locations. Some locations will also require an Explosion Proof (EXP) motor as required by the NEC (National Electric Code) and/or the NFPA (National Fire Protection Association). *Much safer.*

Safer Working Environment

Collecting and exhausting fumes and smoke "*at the source*", provides a much healthier working environment. *Protects workers.*

Rugged Construction

Heavy-duty construction of all fan components will provide long, dependable service. There are many Cincinnati Fan Fume Exhausters still in operation in small welding shops and factories for over 20 years. *Many years of dependable service.*

Made in the U.s.a.

In an era when small machinery manufacturing is continually going over seas, we are proud that Cincinnati Fan "Fume Exhauster-Blowers" are still made and assembled in our factory in Mason, Ohio. You will find imported products, some of which are near copies of our design and/or claim to be as efficient or deliver more CFM. If you do, first ask them for certified performance data. Ours is in our catalogue. Your selection should not turn out to be a disappointment. Compare the features, benefits and component material. We also offer many accessory nozzles, reducers, hose, clamps, and dollies to fit our units.

Imported models don't offer these additional system components. And, who will you call for assistance or service? *Help support the U.S.A.*

Engineering Assistance

Our catalogue contains information on how to select the proper size fume exhauster. *No guess work.*

Manufacturers of Industrial Fans made in North America

Jan Fan is a leading manufacturer of industrial fans that include: heavy duty pedestal fans, wall & ceiling fans, shop fans, floor fans, and mobile fans.

Our industrial fans are considered to be the most energy efficient, maintenance free fans on the market. We are proud of the fact that we have become the automotive and heavy industry standard, a reputation earned by providing superior fan quality and outstanding customer service for over 50 years. When it comes to industrial fans, Jan Fan is considered the fan of choice throughout North America.

The reasons for our success are many. Our yellow industrial fans are extremely energy efficient and built to last; yet they are competitively priced with fans of far less quality. Each of our industrial fans is engineered for ease of maintenance and has proven its durability over time and under the most rigorous circumstances.

Fan is known as the innovator, not imitator of the industrial fan industry. Over the years having been awarded 8 patents, we have designed and introduced many firsts to the market of industrial fans. These include the revolutionary "Smart Air Circulator", energy saving module that automatically turns the fan off and many other unique cost saving features.

We offer 20", 24" and 30" heavy duty industrial air circulators and a full range of replacement fan parts and accessories that are interchangeable with all previous models of our design. A variety of unique and fully adjustable mounting options are available for floor fans, wall column, and bench fans. In 2002 we added our new 12" industrial workstation fans to our line.

By far, Jan Fans offer the greatest cost savings of all heavy duty industrial fans on the market. We provide a massive inventory that is monitored to provide SAME DAY shipments if ordered before 2:00PM (Eastern Time). Jan Fan's yellow industrial fans are sold through distributors throughout the United States, Canada and Mexico.

Our heavy duty industrial fans will exceed your expectations. Unless you try a Jan Fan, you will never know what you are missing because...

Industrial Duty Centrifugal Fans

The Industrial Duty Centrifugal Fans, series 41, are designed for heavy duty air movement applications. Series 41 construction includes both backward inclined and airfoil centrifugal wheels fans and features heavy gauge, fully welded steel housings on all sizes. Stainless steel and aluminium construction materials are available for either the entire unit or air stream parts.

Operated in broad range of fan applications, typically in ducted systems. Versatile construction options allow use in environments which require spark resistance, high temperature tolerance, or resistance to corrosive elements.

Models BIDW/AFDW: Double Width

Operated in non-ducted inlet applications, primarily handling clean air at ambient temperatures. Higher volume capacities allow for a more compact system design than with single width fans.

Greenheck's Total Centrifugal Fan Package

Greenheck offers a complete fan package including your choice of

- Motors and Drives
- Mounting Bases and Vibration Isolation
- Dampers for Centrifugal Fans
- Protective Coatings
- Safety Guards and Weatherhoods.

The advantages of buying a complete fan package from Greenheck include proper selection, design, testing and manufacturing.

Industrial Heavy Duty Fans

Heavy Duty Fans for Industry

Fans play an important role in influencing the operation and overall economy of many process plants.

Flakt Woods extensive experience in the application of centrifugal and axial fans of all sizes means a design to suit the performance characteristics and material of construction appropriate to each installation requirement.

Both in Australia and worldwide, Flakt Woods has established a name synonymous with quality and innovation for the design and supply of fans for:

- Power Industry
- Cement Industry
- Minerals Processing Applications
- Gas Cleaning Systems
- Boiler Installations
- Aeration/Flotation
- Pulp and Paper
- Petrochemical
- Food industries
- Water Treatment
- Mechanical Vapour Recompression
- General Process Fans.

Strong local knowledge combined with extensive international contacts ensures we can design, supply and install fans for almost any industrial application.

Blowers/Fans/Filters

The worldwide market for tube-axial fans is very large and, as a consequence, the number of tube-axial suppliers is significant. This article will focus on the techniques that can be used to quickly compare various tube-axial fan product offerings, in order to provide a quick screen and minimise the number of suppliers that are submitted to a detailed evaluation.

Supplier Classification

In simplistic terms, fan suppliers can be categorised as suppliers of: Low-Cost Fans, Competitive Fans, and Fans Sold Through Distribution.

Low-Cost Fans: Here, price is the main driver in determining the fan design, as well as the manufacturing process. Fans in this category generally utilise techniques that minimise cost. Plastic parts, when viewed under magnification, will have a high degree of tool marks present, indicative of a low degree of tool polishing (minimising tool cost). These fans will also have a higher amount of parting line or ejector flash on their plastic parts.

The circuit board design is simple and is dominated by low-cost components, with the final assembly having a large degree of solder flux present. Motor control or signal output options are typically not available.

Competitive Fans: This category is dominated by major electronic manufacturers, which typically build to order and may have minimum order requirements.

These fans utilise a more complex board design, incorporating various options. A third wire may be included to provide some type of signal output, such as fan rpm or if a locked rotor occurs. The circuit board design typically utilises an integrated circuit as the main motor controller.

Complete solder flux removal on the circuit board assembly is apparent. Plastic parts exhibit a high degree of tool polishing, as well as minimal parting line and ejector flash. In addition, more emphasis on aerodynamics in designing the frame and impeller is present (this leads to slightly more expensive tooling).

Fan Through Distribution: Fan design is similar to the Competitive Fans class. Manufacturers whose strategy is based on a large distribution focus dominate this category. As such, they concentrate on lower volume applications, where the distribution strategy is an advantage. Typically, these will be associated with a slightly higher price structure with no minimum order requirements.

Comparison Techniques

Fan testing is expensive and time consuming, which makes it worthwhile utilising quick comparison techniques in order to identify the few suppliers that require follow-up evaluation. The following list identifies simple comparison techniques, with no particular emphasis on order.

Fan Pricing

Ball bearing designs are more expensive than sleeve bearings and, as a general rule, can add an extra dollar to the price. The issue of ball versus sleeve bearings is application dependent and will not be discussed here, as a multitude of articles have been written on the subject.

Fan Design

The method with which the impeller is attached to the frame is crucial, in order to ensure that the impeller does not come loose when exposed to shock conditions. Most designs utilise metal retention rings, which snap into a groove on the impeller shaft. Disassembly of the impeller from the frame provides a quick insight into the robustness of the fastening method of the impeller shaft to the frame.

With the impeller off, the stator/bobbin assembly is now accessible. Some designs utilise stacked lamina sprayed with a dielectric paint (called fluidizing), that enables it to provide a dual function as a bobbin. There are two inherent advantages with this approach. As the stacked lamina is coated, it is able to withstand a harsher environment before the steel lamina starts experiencing severe rusting. The other is that it provides the appropriate housing support, if a ball bearing design is used.

The main disadvantage is cost and the increased potential for electrical shorts. This design is more expensive to manufacture and, if the dielectric is missing or weak, shorting of the windings can occur. Other designs utilise a plastic bobbin that has the stacked steel lamina attached to the periphery. As the bobbin is plastic, shorting of the coils is minimised. In fan designs incorporating ball bearings, the bobbin must also be able to provide the appropriate circumferential support to the ball bearing outer race. Bearing manufacturers usually specify a press-fit interference between the outer race of the bearing and the inner diametre of the housing. This is required in order to ensure the proper radial play, internal to the bearing, is present. The proper press-fit interference will result in a preferred running clearance with minimal radial run-out. Both of these are major factors in ensuring the maximum life of the ball bearing and, at the same time, provide a smooth and quiet operation.

Next, we examine the circuit board for any solder flux residual. Removal of the circuit board will determine whether it is a one or two-sided design and the type of component make-up (surface mount, thru-hole or a combination of both). The most expensive designs to manufacture are those that combine both surface mount and thru-hole technologies. Long-term availability of some thru-hole components may be an issue. Designs that are all surface mount on one side and with only the bobbin/stator assembly on the other, may have a manufacturing advantage, be it cost and/or quality.

Investigate if the fan has polarity protection, as the consequences for not having this are obvious (mis-wire the fan during the application and the results are disastrous).

Reliability

An understanding of the standard qualification testing can give an insight into the design robustness of a fan. Typically, the fan is subjected to some of the following types of environments: vibration, shock, high and low temperatures, humidity, and/or a combination. An ongoing life-testing program will provide a better picture of the long-term reliability of the design. Since there is no universal standard for fan life and qualification testing, a description of the environmental test conditions is required, as well as the criteria used to determine if failures occurred. A good way to judge a supplier's reliability program is to ask for actual test data, both on the qualification of the design as well as actual life testing.

Current Control Fans

Control of cooling fans has become increasingly popular as efforts are made to reduce system power consumption and mechanical fan noise. The purpose of including a fan is to reduce temperature and keep it as low as possible. Mean time to failure for a variety of systems increases rapidly as temperature elevates.

This has brought about a wide choice of thermally based fan control IC's. Using temperature to control a fan in fact conflicts with the ideal of keeping systems as cool as possible.

Thermal fan control requires a temperature rise to enable the fan as depicte. Proportional controllers make fan speed proportional to temperature, which is even more inappropriate, since it requires even higher temperatures to enable higher fan speeds. Such control strategy is more appropriate in a system where it is desired to stabilise a system at an elevated temperature, rather than to cool it as much as possible.

Fan speed should be based on the thermal load of the system, which is power dissipation. In most electronic systems, power dissipation is directly proportional to power consumption, which is readily measured as the power supply current into the system. In addition to being a true measure of the thermal load, supply current is an instantaneous indicator of thermal load, revealing the need for additional cooling long before the temperature of even on-chip silicon diodes can detect it. Essentially, fan speed should be based on system power supply current as depict.

Computers, both desktop and notebook, all require fans and associated control systems. At this time fan control is accomplished thermally. Certainly in computers sensing CPU current can be used to control fans, and this would provide lower operating temperatures than existing methods of fan control. In computers where the manufacturer prefers "stand alone" fan control without software involvement, current sensing is the optimum solution. Computers where software is already involved on a continuous basis with fan control can take advantage of software input to fan control from computer performance monitors and tools that keep track of CPU activity. Fan speed in such a system would be proportional to that CPU activity.

A fan control scheme based on temperature that requires a rise in temperature to act. The thermometre has to go up to turn the fan on and make it go faster. This is not a maximum cooling scenario. A fan control scheme in which fan speed is controlled by sensing supply current. This makes fan control proportional to the thermal load, eliminating time lags and insuring the lowest possible system temperatures.

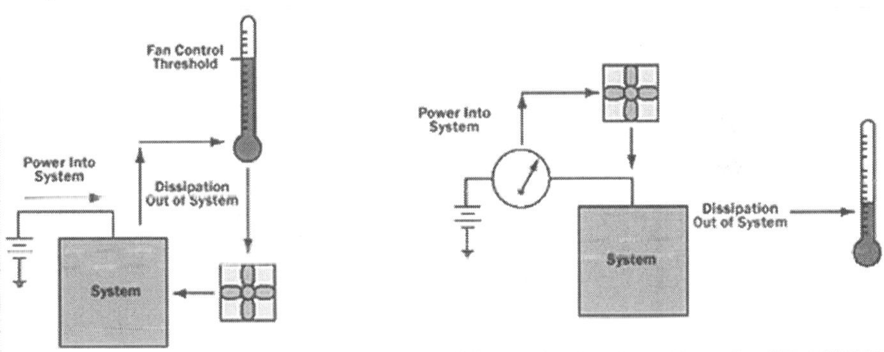

Figure : (a) Fan speed control proportional to temperature. (b) Fan speed control proportional to thermal load.

To further illustrate the benefit of fan control by current measurement, compare the graphs using the example of a computer CPU. The graph shows the power supply current and depicts CPU temperature behaviour with a thermally controlled fan and a current controlled fan. The current controlled fan reacts immediately to the thermal load where the thermally controlled fan must wait until temperature rises. The current controlled fan will always run at full speed if full current is being drawn regardless of temperature, insuring the lowest possible operating temperature.

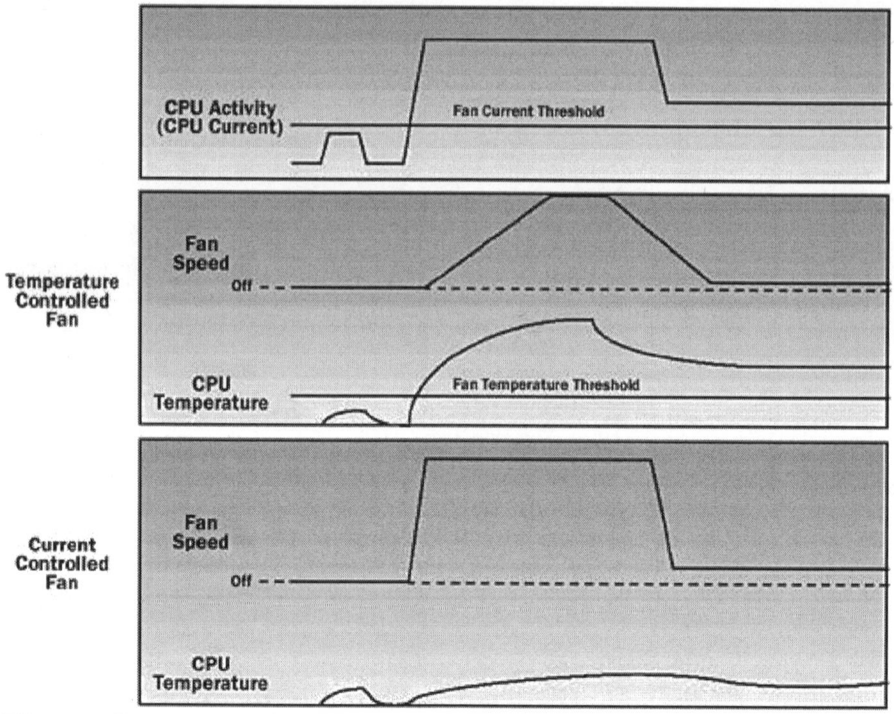

Figure : Comparison of CPU temperature behaviour for thermally controlled fan vs. current controlled fan. By reacting faster and harder, the current controlled fan prevents large temperature excursions and keeps the CPU as cool as possible.

While this is simple enough on the surface, power supply current varies dynamically and rapidly far more than the natural integration that occurs in temperature measurement. Practical current controlled fan drivers will incorporate integration or delays, in the case of analogue circuitry, or algorithms, in the case of processor control, to smooth the fan response.

Simplify implementing current controlled fan designs by using dedicated high-side current shunt monitoring ICs as shown in the CPU sensing. IC1 senses CPU current in RS, which is sized for a maximum of a 100mV drop at full rated CPU current. IC1 converts that drop to a current at IC12 s output (200ýÿA/V) with R1 providing a load to develop a voltage output. IC12 s output voltage can only swing to a maximum of 1.7 volts on a 2.7 volts supply determining the value for R1 based on a maximum input of 100mV and IC12 s scale factor. A1 provides the necessary gain *(Gain* = 12/1.7 = 7) to drive a 12-volt fan from the 1.7 volt output swing of IC1. Q1 buffers A12 s output to provide sufficient current to drive the fan. This very simple realisation uses only integration via C1 to smooth the fan response to current fluctuations, with the tradeoff of slowing the fan response. (Arguably the fan response is still much faster than thermal sensing, besides the fact that it forces the fan speed to a setting appropriate to reducing temperature). C1 would normally be found empirically after evaluating the thermal performance of the target system.

Blowers/Fans/Filters

There are three basics to creating low-noise axial fan applications:

- airflow path design
- initial fan design
- analysis and correction of specific acoustical responses in the physical model

Each is important and contributes to the overall sound levels of the program.

The final step is confirming that the proper flow is achieved and then analysing the fan's noise signature on a decibel-versus-frequency graph. This review allows the designer to find and correct noise "spikes", such as blade pass, imbalance and natural frequencies. While this last step is often overlooked, it can help greatly, especially if modification can be made to the design that was created in step two above.

An earlier work by the authors (Molki et. al., 1995) discussed the adiabatic heat transfer coefficient of rectangular blocks situated in an in-line array. In real circuit boards, heat is generated by a number of electronic components. The temperature of each component is affected by two factors, self-heating and thermal wake effect. Self-heating is the temperature rise due to heat generated within the component, while thermal wake effect is the heating caused by the thermal wake of the upstream heated components. The purpose of this article is to indicate how the adiabatic heat transfer coefficient can be employed to predict the temperature of the components simulated by an in-line array of rectangular blocks with random heating. The focus of the work is on the entrance heat transfer coefficient, which has more practical application. Pressure drop results have also been obtained and reported elsewhere (Molki et al., 1994).

Review of the Literature

Heat transfer in arrays of rectangular blocks has been addressed by a number of investigators. Sparrow et al. (1982, 1983) reported adiabatic heat transfer coefficients in arrays of rectangular blocks with barriers and missing blocks. Arvizu and Moffat (1982) discussed a superposition approach for predicting the temperature of a regular array of cubical elements. Wirtz and Dykshoorn (1984) studied heat transfer in a sparse array of elements. In a series of experimental investigations, Moffat and his group have extensively studied thermal-hydraulic behaviour of the arrays of electronic components (Moffat et al., 1985, Moffat and Anderson, 1998, Anderson and Moffat, 1990, 1991). Other relevant studies are those reported by Faghri et al. (1991), a review by Peterson and Ortega (1990), Anderson (1997), and Leung and Kang (1998).

The Experimental Setup and Procedure

The experimental setup consists of a bell-mouth inlet, a flow development section (591 mm), a test section (343 mm), a flow redevelopment section (591 mm), a venturimetre, a control valve, and a blower. The copper rectangular blocks are positioned along the lower wall of the test section in an in-line arrangement. The geometry of the test section is identified with S, H, L, and B. In this study, the range of parameters was S = 6.4 – 8.4 mm, H = 6.4 – 25.4 mm, L = 25.4 – 49.8 mm, and B = 12.7 – 25.4 mm, corresponding to S/L = 0.128 – 0.33, H/L = 0.128 – 1, and B/L = 0.5.

In a typical data run, only one copper block was heated at a time, and the remaining blocks were adiabatic. The input power was adjusted such that the temperature difference between the block and ambient remained constant (30° to 50°C). The convective heat transfer coefficient is evaluated from

$$h = \frac{\dot{Q}}{A(T_h - T_{in})} \quad (1)$$

where $\dot{Q}$ is calculated as heat generated by the heater, V^2/R, minus the conduction and radiation heat losses. Conduction losses were evaluated from a numerical simulation of heat transfer in the copper block and the adjacent walls. Radiation losses were estimated from a simplified model where the heated block is treated as a small radiating object surrounded by a large environment. It was found that conduction and radiation losses were 10% and 1% of V^2/R, respectively.

Temperature Prediction

The data reported earlier by the authors (Molki et al., 1995) included details of the adiabatic heat transfer coefficients in the entrance and fully-developed regions of the in-line array of rectangular blocks. The focus of this article, however, is on correlations obtained from the earlier data.

Adiabatic Heat Transfer Coefficient

The heat transfer coefficient obtained from Eq. (1) is the *adiabatic heat transfer coefficient*. Based on the present experimental data, the fully-developed adiabatic heat transfer coefficient is obtained from

$$Nu_{fd} = 0.343Re^{0.607} (S/L)^{0.295} (H/L)^{-0.670} \quad (2)$$

where Nu_{fd} =hL/K. The entrance (developing) adiabatic heat transfer coefficient is a function of the axial distance, X, and is expressed as

$$Nu/Nu_{fd} = 1+0.0786(X/D_h)^{-1.099} \quad (3)$$

Equations (2) – (3) are suggested as a complete set of correlations for the adiabatic heat transfer coefficient.

Adiabatic Temperatures

Temperature of the heated block (when only one block is heated), T_h, is obtained from Eq. (1). The temperature of the adiabatic block situated immediately downstream of the heated blocks is obtained from

$$\theta_1=0.372Re^{0.168}(S/L)^{-0.540} \quad (4)$$

where $\theta 1=(T_{a1}-T_{in})/(T_h-T_{in})$. Temperatures of the 2nd and subsequent blocks situated downstream of the heated block are expressed as

$$\theta_N/\theta_1=0.151+0.849N^{-1.314} \quad (5)$$

In this equation, N equals 2, 3, ..., for 2nd, 3rd, ..., blocks situated downstream of the heated block, respectively. These downstream adiabatic blocks are sitting in the thermal wake of the heated block and are heated by the heat generated upstream. The blocks in the flanking columns are also affected by the thermal wake of the heated block. The *wake effect* for the flanking column may be represented by

The worldwide market for tube-axial fans is very large and, as a consequence, the number of tube-axial suppliers is significant. This article will focus on the techniques that can be used to quickly compare various tube-axial fan product offerings, in order to provide a quick screen and minimise the number of suppliers that are submitted to a detailed evaluation.

Supplier Classification

In simplistic terms, fan suppliers can be categorised as suppliers of: Low-Cost Fans, Competitive Fans, and Fans Sold Through Distribution.

Low-Cost Fans: Here, price is the main driver in determining the fan design, as well as the manufacturing process. Fans in this category generally utilise techniques that minimise cost. Plastic parts, when viewed under magnification, will have a high degree of tool marks present, indicative of a low degree of tool polishing (minimising tool cost). These fans will also have a higher amount of parting line or ejector flash on their plastic parts. The circuit board design is simple and is dominated by low-cost components, with the final assembly having a large degree of solder flux present. Motor control or signal output options are typically not available.

Competitive Fans: This category is dominated by major electronic manufacturers, which typically build to order and may have minimum order requirements.

These fans utilise a more complex board design, incorporating various options. A third wire may be included to provide some type of signal output, such as fan rpm or if a locked rotor occurs. The circuit board design typically utilises an integrated circuit as the main motor controller. Complete solder flux removal on the circuit board assembly is apparent. Plastic parts exhibit a high degree of tool polishing, as well as minimal parting line and ejector flash. In addition, more emphasis on aerodynamics in designing the frame and impeller is present (this leads to slightly more expensive tooling).

Fan Through Distribution: Fan design is similar to the Competitive Fans class. Manufacturers whose strategy is based on a large distribution focus dominate this category. As such, they concentrate on lower volume applications, where the distribution strategy is an advantage. Typically, these will be associated with a slightly higher price structure with no minimum order requirements.

Comparison Techniques

Fan testing is expensive and time consuming, which makes it worthwhile utilising quick comparison techniques in order to identify the few suppliers that require follow-up evaluation. The following list identifies simple comparison techniques, with no particular emphasis on order.

Fan Pricing

Ball bearing designs are more expensive than sleeve bearings and, as a general rule, can add an extra dollar to the price. The issue of ball versus sleeve bearings is application dependent and will not be discussed here, as a multitude of articles have been written on the subject.

Fan Design

The method with which the impeller is attached to the frame is crucial, in order to ensure that the impeller does not come loose when exposed to shock conditions. Most designs utilise metal retention rings, which snap into a groove on the impeller shaft. Disassembly of the impeller from the frame provides a quick insight into the robustness of the fastening method of the impeller shaft to the frame.

With the impeller off, the stator/bobbin assembly is now accessible. Some designs utilise stacked lamina sprayed with a dielectric paint (called fluidizing), that enables it to provide a dual function as a bobbin. There are two inherent advantages with this approach.

As the stacked lamina is coated, it is able to withstand a harsher environment before the steel lamina starts experiencing severe rusting. The other is that it provides the appropriate housing support, if a ball bearing design is used. The main disadvantage is cost and the increased potential for electrical shorts.

This design is more expensive to manufacture and, if the dielectric is missing or weak, shorting of the windings can occur. Other designs utilise a plastic bobbin that has the stacked steel lamina attached to the periphery. As the bobbin is plastic, shorting of the coils is minimised. In fan designs incorporating ball bearings, the bobbin must also be able to provide the appropriate circumferential support to the ball bearing outer race. Bearing manufacturers usually specify a press-fit interference between the outer race of the bearing and the inner diametre of the housing. This is required in order to ensure the proper radial play, internal to the bearing, is present. The proper press-fit interference will result in a preferred running clearance with minimal radial run-out. Both of these are major factors in ensuring the maximum life of the ball bearing and, at the same time, provide a smooth and quiet operation.

Next, we examine the circuit board for any solder flux residual. Removal of the circuit board will determine whether it is a one or two-sided design and the type of component make-up (surface mount, thru-hole or a combination of both).

The most expensive designs to manufacture are those that combine both surface mount and thru-hole technologies. Long-term availability of some thru-hole components may be an issue. Designs that are all surface mount on one side and with only the bobbin/stator assembly on the other, may have a manufacturing advantage, be it cost and/ or quality.

Investigate if the fan has polarity protection, as the consequences for not having this are obvious (mis-wire the fan during the application and the results are disastrous).

Reliability

An understanding of the standard qualification testing can give an insight into the design robustness of a fan. Typically, the fan is subjected to some of the following types of environments: vibration, shock, high and low temperatures, humidity, and/or a combination. An ongoing life-testing program will provide a better picture of the long-term reliability of the design.

Since there is no universal standard for fan life and qualification testing, a description of the environmental test conditions is required, as well as the criteria used to determine if failures occurred. A good way to judge a supplier's reliability program is to ask for actual test data, both on the qualification of the design as well as actual life testing.

New 'fan' Design has no Moving Parts

Aside from hard drives, one of the main culprits behind noise production in today's PCs is air cooling. Fans have definitely gotten bigger and quieter in recent years, but they can still be loud—especially if they start to wear out and begin ticking or humming. According to iTnews Australia, however, U.S.-based researchers have developed a new "fan" design that doesn't actually have any moving parts.

The design is nicknamed RSD5, and iTnews Australia says it took Dan Schlitz and Vishal Singhal of Thorrn Micro Technologies six years to develop. In short, RSD5 uses electric fields to generate a plasma that can move air. Singhal describes it as follows:

He explained that RSD5 incorporates a series of live wires that generate a micro-scale plasma (an ion-rich gas that has free electrons that conduct electricity).

The wires lie within uncharged conducting plates that are contoured into half-cylindrical shapes to partially envelop the wires.

Within the intense electric field that results, ions push neutral air molecules from the wire to the plate, generating a wind. The phenomenon is called corona wind.

The result is a device that generates three times the air flow of a small fan at just a quarter of the size. In fact, iTnews says an RSD5 "smaller than one cubic-cm" is capable of cooling a 25W chip—roughly the power envelope of today's laptop processors. Singhal and Schlitz believe they can go so far as to integrate the cooling system into the silicon to make "self-cooling chips," as well.

New, Patented Technology By Supermicro Creates First Intelligent Cooling Fan Backup For Power Supplies

A patented new redundant cooling system that provides the market's first intelligent cooling fan back-up for power supplies has been introduced by SUPERMICRO of San Jose, CA.

This new system intelligently detects and responds to increases in temperature that result from the breakdown of a simple fan, the weak link in any power supply system.

"A reliable, dependable power supply is crucial to the stability and performance of your system," said Charles Liang, president of SUPERMICRO. "This innovative technology prevents the disasters that can happen when the cooling fan breaks down and the power supply is interrupted."

The redundant cooling system includes an additional back-up fan, which operates independently from the primary fan, and a thermal control circuit that includes a solid state temperature sensor. This sensor detects ambient temperature changes inside the power supply, and is connected to the fan power drives of both the primary and back-up fans. The thermal control circuit works with the temperature sensor to trigger a response whenever specific high and low temperature thresholds are exceeded. This circuitry enables the primary fan to increase its RPM simultaneously with increases in the ambient temperature until reaching full speed at 40°C.

Power supply fans inevitably break down over long periods of use, due simply to the wearing out of their parts. If this happens, or if the ambient temperature exceeds approximately 68°C, the cooling system's back-up fan automatically turns on, an alarm sounds and a warning LED is illuminated.

The power supply includes an external button to turn the alarm off, but the LED display remains lit until corrective measures have been taken.

These warning signals give system administrators advance notice of fan breakdown or dangerous rises in the power supply temperature. In addition, the system will be protected until it can be safely shut down to replace the primary fan or to correct the cause of the heat problems.

These Redundant Cooling Power Systems are especially appropriate for servers or workstations, where system integrity is critical.

The Redundant Cooling Power Supply is available in 400 W and 300 W models.

Redundant Cooling Power Supply Specifications

- 400 W model • P/N SP401-RA
- 300 W model • P/N SP301-RA
- 150 x 86 x 170 mm (WxHxL)
- +5V, -5V, +12V, -12V and +5V standby outputs
- UL, CSA, TUV and CE certified

SUPERMICRO is a leading supplier of high-end motherboards and servers. Since its founding in 1993, SUPERMICRO has led the industry in delivering cutting-edge technology such as the first solution for the I2O architecture. SUPERMICRO strives to build long-term relationships with its customers by providing outstanding engineering innovations, quality and reliability. SUPERMICRO's server building block solutions decrease systems integrators' time-to-market by ensuring the interoperability of all components. All SUPERMICRO products are available through OEMs and distributors.

Clever New Fan has No Blades

Few inventions have lent themselves to as little improvement over the years as the simple fan, with its whirring blades that blow air across a desk or through a room. But now even the conventional fan has been transformed.

James Dyson, the British inventor behind the vacuums that bear his name, has come up with an incredibly simple-looking, circular fan that has no blades. It uses "Air Multiplier" technology to push 119 gallons of smooth air out every second.

The conventional fan relying on blades has been largely unchanged for about two centuries; the first mechanical fans are said to have been built in the Middle East in the early 19th century.

Ceiling fans blew onto the scene in the 1860s, powered by running water and a setup that turned a series of belts with the ability to operate several fans at once (some of these still exist in old buildings). A personal electric fan hit the market in the 1880s.

But the traditional design is not perfect, Dyson figured. Fan blades chop the air, creating an uneven airflow with buffeting bursts and moments of weaker flow.

Here's how the new setup works:

Air is pulled into the machine's cylindrical base with an impeller that draws from jet-engine technology. The air rushes up into a hollow ring. The air is then forced out of a 1.3-millimetre-wide slit (less than 1/16th of an inch) that runs all around the ring.

Here's the tricky part, developed during four years of testing:

As the air exits through the slit in the ring, it flows over the inner edge of the ring, which was modelled after an airplane wing. Imagine an airplane wing bent around into a loop.

The air encounters what would be the leading edge of the wing, said Dyson engineer Andy Samways. When it hits the wing-like structure, the air speeds up, creating an area of low pressure.

"Air or any fluid will tend to hug a smooth surface," Samways explained in a telephone interview. "Air over a wing follows the profile, and that creates lift."

As the air exits the loop, the lower pressure pulls air from behind the fan along with it, and air around the front of the fan also gets pulled into the stream. That's simple physics. But it was the wing-like ramp that served as a breakthrough.

"We realised that this inducement, or amplification, effect could be further enhanced by passing airflow over a ramp," said James Dyson. "This was the point where the idea of a bladeless fan became a real possibility. Here was a way to create turbulent-free air and finally do away with blades."

Getting rid of blades has visible benefits: The new fan needs no safety grill, and what few parts there are can be easily cleaned. It runs on a small motor, and airflow can be fully regulated with a dimmer switch.

Chapter 9

New Fan Technology

The engineering team at temperzone are specialists in the application of advanced world technologies into the custom built reverse cycle packaged air conditioning units up to and beyond 180 kW cooling capacity. The innovation in temperzone has led this company to the forefront in the development of energy efficiency due to their expertise in new refrigerants and new air moving fans and motors.

The fan motors that stand out above the others in this respect are the EC motors combined with the backward curved single inlet centrifugal plug fans.

Fan manufacturers ebm-papst and Rosenberg have independently developed their own range of this type of motor/fan combination. Temperzone have applied this new type of motor/fan combination in its 90, 100 and 115kW customised rooftop units named PA 3100, PA 3600 and PA 3800 respectively (only available in New Zealand).

Four fans are needed to move the air in the larger range of 125, 145, and 180 kW called the PA 4100, PA 4600 and PA 4800 series.

Costs and delivery times have been the major hurdles in the introduction of the EC motors and fans into the New Zealand market.

Temperzone has negotiated with the suppliers to significantly raise the Australia and NZ stock levels for the most frequently used motors.

At the end of 2006, temperzone sold the EC motor concept design in the PA 4800 to a client who was willing to try out the practical advantages of the EC motors with backward curved fans and accept the financial penalty on the unit by offsetting this with the savings in speed control and noise attenuation.

All parties, including the client, the final user, all visiting professional engineering consultants and temperzone management were impressed with the final product. This has resulted in an escalating quantity of orders to temperzone from clients and consultants for PA units with EC motors and backward curved fans.

The general consensus of the major advantages of these units with EC motor fans has been:

- Integrated variable speed control from 0 to full RPM by 0–10 Vdc signal.
- Soft starting of motors gives no inrush current.
- Efficiency between 95–98% compared to 50–70% for belt drive centrifugal fans.
- Supply air unit outlet size is the same height as the inlet size, thus reducing the outlet velocities from 18–20 m/s to 8–10 m/s and so eliminating the need for tapered ducting transitions.
- Supply air sound power levels are significantly lower due to lower unit exit velocities, removal of belt/drive noise and accoustic attenuation being possible in the supply air pressure chamber of the unit.
- Simplified unit design due to supply air pressure chamber within the unit and the possibility to locate the supply air spigot in many positions without altering the location of the fans.
- Fixed and stable airflows via differential pressure transducer and controller to compensate for varying duct static pressures caused by dirty filters and modulating dampers.
- Commissioning and maintenance time and costs are reduced by eliminating pulley and belt adjustments, wear and changes.

New Technology from Casablanca Fan Company

Casablanca Fan Company has stepped up their control technology for three ceiling fan models the Bel Air Fan, the Bel Air Halo fan and the Mission fan. This soft launch of the Power Touch control will happen during the Summer of 2007. Casablanca fans will release the full launch converting all Inteli-Touch controls to Power Touch controls.

What makes the Power Touch Control Better?

The Power Touch control uses power line communications that can not be effected by outside interference such as garage door openers and wireless networks. Normal radio frequency control are usually effected. There is also no antenna need for operation of this control.

Power line is proven technology used in utility products available from GE, ABB, Dimat and Hunt. Power Touch uses existing wiring to communicate with the desired ceiling fan. Power Touch is also immune to signal attenuation caused by walls, metal objects and people unlike the radio frequency controls.

The Power Touch Control is Easier to Use

Casablanca ceiling fans are about to experience easier operation by having a more simple interface than the inteli touch wall control. The Power Touch control will still have all of the inteli touch features as a base. This control will also have a more contemporary design. These Power Touch controls can operate multiple ceiling fans on one control and they will be compatable with whole house dimming systems.

Prevention of Compressor Short Cycling in Direct-expansion (DX) Rooftop Units : Theoretical Analysis and Simulation

Packaged RTUs offer more advantages than the conventional central system, such as greater flexibility, lower installation cost, and less space use. They have been widely applied in low-rise commercial buildings as well as industrial facilities. However, compressor short cycling, exaggerated by recent wider application of variable air volume (VAV) systems, has not yet been successfully resolved with satisfactory partial load performance and minimum initial cost. It is believed to be the most common cause of early maintenance problems and compressor failures. In addition, poor partial load efficiency and indoor humidity control are also results of compressor short cycling.

It has been identified and accepted by all the researchers that the root cause of this problem is that system load modulation cannot match building load variation. The air conditioning systems are typically designed for an expected peak load, which might happen on the order of 20 to 30 hours per year. Most of the time, the system load varies over a wide range from 80% to as low as 5% due to lighting, occupancy, weather conditions, etc. variations. As the mismatch between the building load and the cooling output occurs, the compressor short cycling happens. Some major efforts that intend to resolve compressor short cycling issues were reviewed in the following sections.

Hot Gas Bypass

Hot gas bypass has long been used in the industry as a refrigeration system capacity control method. It varies the system capacity by introducing an artificial load into the compressor when the system load declines. This control strategy is accomplished through the use of a hot gas bypass valve to bypass high-pressure hot gas from discharge line into the low-pressure side of the refrigeration system (Dossat and Horan 2002). The hot gas bypass valve is calibrated to open when the low-side pressure falls below a preset minimum limit.

Because hot gas bypass provides the continuous system capacity modulation down to 0%, this method indeed can effectively prevent the compressors from short cycling and improve the system partial load performance as well as room comfort. However, intensive attention must be paid to design the bypass system so that a reliable and stable operation can be achieved.

Moreover, its inherent lower efficiency operation under partial load conditions might be another concern. Solberg and Bradley (2003) demonstrated that the risk of oil and refrigerant logging is also inherent to hot gas bypass. We cannot simply assume that adding hot gas bypass will improve reliability or enhance capacity modulation. None of the benefits of the hot gas bypass will be achieved without meticulous attention to design and implementation, which normally cannot be guaranteed for every application. Therefore, they suggested that users only implement hot gas bypass if all other design options fail to meet the demand of the application and leave the hot gas bypass as the last resort.

Multistage Control

The simplest form of capacity control is on-off cycling of the compressor. With such a method, the unit either runs at full capacity or not at all. When under low load condition, this could lead to poor room comfort control and reduce the life expectancy of the compressor due to short cycling. To prevent excessive cycling, a minimum compressor on/off time might often be integrated with this method. This method is typically employed in the small tonnage RTUs with constant volume (CV) operation and is no longer employed in the medium to large size RTUs. For the larger RTUs, multiple compressors are used to achieve better turndown and resolution of the capacity modulation. The compressors will sequence the operation by cutting on/off to match the load. Cylinder unloading sometimes may be integrated with multiple reciprocating compressors to reduce the number of compressors while still realising the same turndown and resolution.

This method was the most popular capacity modulation applied in current DX RTUs industry. Based on the applications, the commercial RTU systems available in the U.S. market can be categorised into two groups: 1) serving low-rise buildings, manufacturing/distribution facilities and retail outlets, typically sized less than 30 ton (105.5 kW); 2) serving large office building and commercial/industrial facilities, typically sized from 30 tons (105.5 kW) to 150 tons (527.5 kW). The small tonnage unit typically uses one or two cooling stages, while the large unit usually utilises up to eight cooling stages. The more stages of this control scheme, the better system performance can be expected. However, the cost and complexity will also substantially increase. Therefore, two to six compressors are typically used in RTUs with a capacity control up to eight stages. With this method, the system operation will be improved as well as the fluctuations of the supply air temperature (SAT) and room comfort, but the gap still exists between each stage as well as the cycling of the compressors.

Variable Refrigerant Volume (VRV) Technology

The VRV technology is accomplished by regulating the speed of the compressor motor. It has been the target of a number of researchers and engineers for many years and is considered as the most efficient capacity control method for refrigeration systems and is expected to keep the compressors from short cycling. However, its application has been limited by the minimum refrigerant vapour velocity requirement to ensure oil return, proper splash system lubrication, and cooling of the motors. Both reciprocating compressors and scroll compressors have the challenges to exceed a 4:1 speed ratio. Though the VRV technique will be a promising solution for the new product, the high retrofitting cost and increased complexity still prevent the VRV technology from being employed into existing units.

With the recent development of variable frequency drive (VFD), the compressor technique, and the tremendous reduction on the cost, manufacturers have been encouraged to apply VRV technology into commercial RTUs.

Other Methods and Research

With the development of direct digital control (DDC) technology, the anti-cycle timer has been widely applied in the industry to prevent the compressors from short cycling. It is accomplished by integrating a minimum on/off time with the compressor control, which is typically three to five minutes. However, this is an open-loop method and prevents short cycling by sacrificing the system performance, e.g. loss in set-point accuracy. Therefore, this method typically works with other capacity control algorithms to achieve better system performance.

A practical method, not found in any published papers, is to lower the economizer lockout temperature so that whenever the DX RTU is running, the economizer is disabled. It seems to the facility operator that the DX RTU runs more "continuously" with this method applied. However, there are two major drawbacks associated with this method. First, because of reduced economizer period during moderate weather condition, there is significant cooling energy waste. Second, for the multiple compressors DX RTUs with four or more step capacity control, although the least stage may be prevented from short cycling, but other stages may still be short-cycling.

A very (1996) developed a control method that could minimise the problem of inability to match the cooling capacity of the refrigeration plant with the varying thermal load in the DX VAV RTUs. This method used several small DX coils with face dampers that would be closed when a coil was deactivated. These small DX coils were connected with the same number of independent condensing units so that the load matching problem could be minimised through staging cooling capacity control. However, such a method is equivalent to distributing the load evenly into several smaller DX systems. Therefore, its initial cost is high and its overall energy efficiency remained a concern. In addition, its control system is much more complicated, and its reliability could not be guaranteed. Moreover, the problem of the control method not matching the capacity still exists between each stage.

Summary

Recently the industry has come up with some major achievements towards solving compressor short cycling, such as hot gas bypass, anti-cycle timer, multiple compressors (cylinder unloading for reciprocating compressor), and combinations of the above methods. Though these methods can more or less improve performance, none of them can thoroughly solve this problem with good partial load energy performance and without a significant initial cost increase. The VRV technology will be the ultimate solution for the future products but restricted deployment in the existing units. This paper tends to present an advanced control algorithm to provide solution on the current DX VAV RTUs operation issues with less cost and significant system performance improvement.

Foundation Theory

In the vapour-compression refrigeration cycle, heat is transferred from a lower temperature source to a higher temperature sink by undergoing a certain amount of work. The Carnot cycle is the ideal refrigeration cycle operating between two fixed temperatures with the highest system efficiency. The refrigeration cycle designed to approach the Carnot cycle in the actual system is the theoretical vapour-compression cycle. Two theoretical single stage vapour-compressor cycles operating at different evaporating and condensing temperature. Based on the fundamentals of the vapour-compression cycle, evaporating and condensing temperature will determine the refrigeration effect. Higher evaporating temperature and/or lower condensing temperature will increase the refrigeration effect, while lower evaporating temperature and/or higher condensing temperature tend to reduce the refrigeration effect. A normalised compressor cooling output to its nominal capacity versus evaporating temperature for different condensing temperatures using R22. The nominal capacity is rated at 45 [degrees] F (7.2 [degrees] C) evaporating temperature and 130 [degrees] F (54.4 [degrees] C) condensing temperature. The cooling capacity reduces to 65% of its nominal capacity when using 32 [degrees] F (0 [degrees] C) evaporating temperature and 150 [degrees] F (65.6 [degrees] C) condensing temperature. While 50 [degrees] F (10 [degrees] C) evaporating temperature and 90 [degrees] F (32.2 [degrees] C) condensing temperature are used, the cooling capacity increases about 40% compared to its nominal capacity.

Therefore, the refrigeration system cooling output can be modulated by varying the evaporating and condensing temperature. Theoretically, the different combination of the evaporating and condensing temperatures can achieve any cooling capacity output between system maximum cooling output and zero. However, in practice, the evaporating and condensing temperature can only be changed within a certain range due to some mechanism limitations; for example, the maximum and minimum condensing pressure requirements, minimum refrigerant vapour velocity requirement for good oil return, no formation of frost on the DX coil, etc. In the conventional multistage units, the compressors are running intermittently to achieve certain cooling capacity. Integrating this theory with the multistage control for each stage cooling, by varying the evaporating and condensing temperature, the cooling capacity can be extended to a wide range so that smooth even continuous load modulation can be achieved between each stage. As the first attempt to apply this theory, this research will start with varying the evaporating temperature while keeping the conventional condenser control to achieve the capacity control.

Algorithm

A schematic diagram of a multistage VAV DX RTU. The proposed algorithm has been focused on the refrigeration system control to maintain desired SAT. Therefore, the controls of other components, i.e. the duct static pressure control and the return fan speed control, will remain the same as in conventional practice. The following sections will introduce the conventional control algorithm for the refrigeration system and the proposed advanced control algorithm.

Conventional Control Algorithm

The conventional compressor staging algorithm is to step up or down the cooling stages to maintain the SAT set point plus or minus a dead band, typically 4 [degrees] F (2.2 [degrees] C). Meanwhile, an anti-cycling timer, typically five minutes, is also built in to prevent rapid stage cycling. A stage change can occur only (1) after the minimum on/off time has expired, and (2) if the SAT is outside the dead band. For example, if the SAT set point is 55 [degrees] F (12.8 [degrees] F), the cooling stage steps up when the SAT is higher than 57 [degrees] F (13.9 [degrees] C) and the anti-cycling time has been passed and vice versa. The cooling stage up and down cycling performance.

Advanced Control Algorithm

The evaporating temperature interacts with the SAT and supply air flow. When applying the foundation theory, there might be several solutions associated with different cooling stages to match a given building load. The advanced control algorithm intends to find these solutions and pick the one with minimised cooling stage changes and least energy consumption as set point.

This method is accomplished by iteratively solving the system model until the stable solutions for supply air flow and SAT are established. Since the cooling stage change will not allow "jumping", that is, if the current cooling stage is the second stage, the next cooling stage could be either first stage or third stage but not zero stage or fourth stage. Therefore, the SAT set point will only need to calculate for the current cooling stage and its adjacent cooling stages. With application of the principle of mass and energy conservation, assuming the supply air relative humidity (RH) at 90% for wet coil, the SAT and supply air flow associated with the three cooling stages for the next step can be calculated using the iteration process. The three main governing equations are listed below.

Q.sub.sen] = CF [M.sub.N] ([T.sub.ra] - [T.sub.sp, N]) (1)

[E.sub.N] = [Florin] (CF [M.sub.N], [T.sub.oa], [T.sub.mix], R [H.sub.mix]) (2)

[E.sub.N] = 60pa x CF [M.sub.N] ([h.sub.mix] - [h.sub.sp, N]) (3)

where

[Q.sub.sen] = Building sensible cooling load, Btu/h (W)

CF [M.sub.N] = Calculated air volume flow rate at Nth cooling stage to satisfy [Q.sub.sen] cfm ([m.sup.3]/s)

[T.sub.ra], [T.sub.oa], [T.sub.mix] = Return air, outdoor air, and mixed air temperature, [degrees] F ([degrees] C)

[T.sub.sp, N] = Calculated SAT set point at Nth cooling stage to satisfy [Q.sub.sen], [degrees] F ([degrees] C)

[E.sub.N] = Cooling capacity of Nth cooling stage, Btu/h (W)

R [H.sub.mix] = Mixed air relative humidity, %

Pa = Mean air density across the DX coil, lbm/f [t.sup.3]

(kg/[m.sup.3])

[h.sub.mix] = Mixed air enthalpy, Btu/lbm (J/kg)

[h.sub.sp, N] = Air enthalpy associated with [T.sub.sp, N], Btu/1bm (J/kg)

The building sensible cooling load is obtained from the measurements of instant supply airflow, return and supply air temperature. The airflow measurement can be challenging, and a method developed by Liu et al. (2005) is recommended. Outdoor air temperature, mixed air temperature, and mixed air relative humidity are also real-time measured data. Mixed air enthalpy, [h.sub.mix], is calculated from the measurements. The function, [Florin], calculates the cooling output at conditions different from rating condition. It can either be the biquadratic and quadratic curves used in EnergyPlus (USDOE 2010) or the one described in Pang (2008). After these equations were solved, the associated system electric power can be calculated accordingly:

[W.sub.cooling, N] = [f.sub.1] ([T.sub.mix], R [H.sub.mix], [T.sub.ca], CF [M.sub.N]) (4)

[W.sub.sf, N] = [f.sub.2] (CF [M.sub.N]/CF [M.sub.d]) (5)

[W.sub.rf, N] = [W.sub.rf, d] [CF [M.sub.ra, N]/CF [M.sub.ra,d]).sup.3] (6)

[W.sub.tot, N] = [W.sub.cooling, N] + [W.sub.sf, N] + [W.sub.rf, N] (7)

where

[W.sub.cooling, N] = Electric power of compressor and condenser fans at Nth cooling stage, kW

[W.sub.sf, N], [W.sub.rf, N] = Electric power of supply and return air fan at Nth cooling stage, kW

CF [M.sub.d], CF [M.sub.ra,d] = Design supply and return air volume flow rate, cfm ([m.sup.3]/s)

CF [M.sub.ra,N] = Resulting return airflow based on CF [M.sub.N], cfm ([m.sup.3]/s)

The variables [f.sub.1] and [f.sub.2] can refer to the curves used in EnergyPlus (USDOE 2010). The solution with the lowest total electric power [W.sub.tot,N] will be selected as the SAT setpoint.

System Model

A DX RTU steady-state model was developed to simulate its operation and justify the advanced SAT reset algorithm. The model input parameters are building load profile and the outdoor air conditions including dry bulb and relative humidity, and the output parameters are various system performance data, such as supply air condition, the evaporating and condensing temperature, and the compressor power consumption. The sub-components of the steady-state model and how they interact with each other. Detailed components models can refer to Pang 2008.

The on/off cycling loss is a key contributor to the degradation of the refrigeration system performance, which attributes to the dynamic behaviour during compressor startup. In order to investigate the cycling losses with the steady-state model, the cooling capacity provided by the compressor during startup is treated as a first-order system with a single time constant; whereas the electric power of the compressor reaches steady-state instantaneously (Mara et al. 2005; O'Neal et al. 1993; Garde et al. 2002; Henderson et al. 1996). Thus, the system dynamics for a specific cooling stage N at any given time t can be expressed by

[Q.sub.N, e, cyc] = [Q.sub.N, e] (1 - [e.sup.-t/[tau]]) (8)

[Q.sub.N, sen, cyc] = [Q.sub.N, sen] (1- [e.sup.-t/[tau]]) (9)

[W.sub.N, cyc] = [W.sub. N, ss] (10)

where

[Q.sub.N, e, cyc], [Q.sub.N, sen, cyc] = Instantaneous total, sensible cooling capacity for Nth stage, Btu/h (W)

[Q.sub.N, e], [Q.sub.N, sen] = Steady-state total, sensible cooling capacity for Nth stage, Btu/h (W)

[tau] = DX cooling coil time constant, second

[W.sub.N, cyc] = Instantaneous compressor electric power for Nth stage, kW

[W.sub. N, ss] = Steady-state compressor electric power for Nth stage, kW

The system time constant [tau] is typically between 60 and 200 seconds depending on the system size. For the large commercial DX unit, the time constant is around 180 seconds.

When the dew point of the air passing through the DX coil is higher than the DX coil surface temperature, the air undergoes a simultaneous cooling and dehumidification process. Latent capacity degradation occurs when the DX coil cycles on and off because a portion of the moisture condensing on the DX coil surface during the cooling on cycle evaporates back into the airstream during the cooling off cycle (Shirey and Henderson 2004). The impact of this latent degradation phenomenon is not accounted for in this study due to the increased complexity of the model.

Simulation and Results

The objective of the simulation is to compare the system cooling performance between the improved operation and the conventional operation for both a four-stage unit and an eight-stage unit on a typical day. The system daily performance is divided into five minute intervals and treated as steady-state within this period. The steady-state model is then used to simulate the daily dynamic behaviour of the DX RTU by using equations 8 through 10. The system model described above was programmed in the engineering equation solver (EES), and the EnergyPlus was used to simulate the space cooling load under different weather conditions. The space cooling load simulation is decoupled from the system simulation using EES, which means the space cooling load is calculated by assuming the room air temperature at set point and room temperature fluctuation caused by the system is not taken into account.

The hourly outdoor air dry bulb temperature and dew points are then used in EnergyPlus to simulation the zone cooling load for each scenario. The zone cooling loads are normalised to the design cooling loads and regression models that relate the hourly zone cooling load to the time of the day are created. The normalised interior and exterior zone cooling loads for the low load and moderate weather scenario. The resulting cooling loads are then used by the EES to carry out the system simulation.

Simulation

The system is assumed to operate daily from 6 a.m. to 8 p.m. The simulation results of the eight-stage unit performance with low load condition under moderate weather condition were presented below. Other simulation results can refer to Pang (2008).

The system air side performance for the eight-stage, the cooling stage change rates for both cases. The improved case provides smooth compressor operation as opposed to the short cycling of the compressor in the base case. The cooling stage changing rate during the 14-hours period is 8 vs. 70 for the improved case and base case. In the base case, significant short cycling occurs in the morning and evening when the refrigeration load is small due to light occupancy and low outside air temperature. The short cycling has been eliminated in the improved case by resetting the supply air temperature set point higher to fully utilise the free cooling and not turn on the mechanical cooling until a breakpoint is reached. The breakpoint is determined by comparing the total electric power without mechanical cooling to that with the first stage of cooling, where the supply and return air fans power using high supply air temperature exceeds the total electric power of the refrigeration system, the supply and return air fans using low supply air temperature.

Simulation Results Summary

The elimination of the short cycling is accomplished without additional electric power comparing to the base case for the high and medium load scenarios under all three weather conditions. Some electric power saving is expected for the low load scenario under the moderate and dry weather conditions while some electric power penalty is presented for the low load scenario under the humid weather conditions. The main reason for the electric power penalty is that under humid weather conditions, the improved case provides better humidity control in the space than the base case does, which results in more latent load in the improved case. However, the model simplification that neglects the latent degradation may have increasing impacts on the simulation results under humid. Weather conditions as opposed to dry and moderate weather conditions. Future research that investigate the impacts from latent degradation are needed.

The conventional compressor staging algorithm causes compressor short cycling especially when outdoor air temperature is less than 65 [degrees] F (18.3 [degrees] C). This extensive cycling rate can damage the compressor in a very short time. Simulation results show that the proposed advanced control algorithm can effectively eliminate the compressor short-cycling with safe and smooth refrigeration system operation, which will greatly improve compressor reliability and reduce maintenance work and cost significantly.

The energy savings resulting from the improved operation is located mainly at the lower load condition. When cooling load is high or humid air condition, the energy savings is limited. The major benefits of implementing the advanced algorithm are improved the system operational reliability and better indoor humidity control with no extra energy cost. However, it should be notice that the simulation doesn't take into account the fact that condensation on the DX coil surface will reevaporate into the air stream during the compressor off cycle. Therefore, future research is needed to further investigate the system performance under humid weather condition.

Empirical Modelling of a Rolling-piston Compressor Heat Pump for Predictive Control in Low-lift Cooling

Inverter-driven variable-capacity air conditioners, heat pumps, and chillers can provide energy-efficient cooling, particularly at part-load capacity. Varying the capacity of vapour compression systems enables operation at lower pressure ratios, or low lift, which fundamentally improves the coefficient of performance of the system by reducing the required compressor work while providing a similar cooling effect. Conventional and low-lift vapour compression cycles for refrigerant R22. A cycle with a lower condensing temperature, higher evaporating temperature, and/or slower compressor speed (the low-lift cycle) can achieve a similar (or even greater) cooling effect (the area below the evaporation process line) to that of the conventional cycle (the area inside the cycle polygon) with less compressor work.

The use of radiant cooling, precooling of thermal energy storage, and predictive control of compressor speed, condenser flow rate, and evaporator flow rate allow for low-pressure ratio, or low-lift, chiller operation (Armstrong et al. 2009a, 2009b) to meet daily cooling loads in a near-optimal manner. The combination of these strategies will here be called low-lift cooling. Radiant cooling requires moderate chilled-water temperatures, around 60 [degrees]F to 65 [degrees]F (15.5 [degrees]C to 18.3 [degrees]C), rather than about 44 [degrees]F (6.6 [degrees]C) for conventional systems, allowing for higher evaporating temperatures and pressures. Pre-cooling thermal energy storage allows night-time operation of a heat pump or chiller, providing lower condensing temperatures and pressures.

Predictive control of compressor speed, condenser flow rate, and evaporator flow rate based on forecast cooling loads and zone temperature response allow the heat pump or chiller power consumption and coefficient of performance to be optimised over a 24 h period to minimise energy consumption and operating cost. In order to achieve this predictive control, it is necessary to have a model of heat pump or chiller performance that maps control variables and exogenous operating conditions to power consumption, cooling capacity, and system coefficient of performance.

This paper will review the state of knowledge about heat pump and chiller performance empirical modelling, present experimental results evaluating the performance of an air-source heat pump with a rotary piston compressor—particularly under low-lift conditions—and present empirical curve-fit models of the performance of the heat pump spanning a broad range of conditions.

Literature Review

Historically, heat pump and chiller cooling efficiency ratings have focused on efficiency for a single design load, an average over seasonal conditions, or a small set of part loads. These cooling efficiency ratings are represented by coefficient of performance (COP), seasonal energy efficiency ratio (SEER) for small air conditioners and heat pumps, integrated part load value (IPLV) for large chillers, and the more recent integrated energy efficiency ratio (IEER) for variable refrigerant volume systems. These are useful to engineers who are selecting efficient systems. However, under a given set of operating conditions, the actual efficiency of a heat pump or chiller can be very different from these rated values. A more complete model of heat pump and chiller performance is necessary for model-based predictive control of compressor speed, condenser flow, and evaporator flow to achieve energy-efficient low-lift operation subject to different outdoor air temperatures, zone-air or chilled-water temperatures, and cooling loads.

Existing models for heat pumps and chillers fall broadly into two categories: empirical models and physics-based models. Empirical models are typically regression-based polynomial curve-fit models of heat pump or chiller performance, such as the DOE-2 chiller model (DOE 1980), but may also include neural networks and more advanced black-box modelling methods (Swider 2003). Physics-based models vary in complexity and degree of detail, but all apply fundamental physical principles to heat pumps, chillers, and their components to model system performance. Physics-based gray-box models such as the Gordon-[N.sub.g] Universal Chiller Model (Gordon and [N.sub.g] 2000) have parameters that can be identified from measured performance data for a given system, whereas white-box models use known engineering quantities to model performance.

Jin and Spitler (2002) and Sreedharan and Haves (2001) provide a broad review of heat pump and chiller models focused primarily on steady-state chiller performance. Benapudi and Braun (2002) and Rasmussen (2005) review transient models of heat pumps and chillers more suitable for model-based transient control optimisation or to investigate new heat pump or chiller design options. The steady-state performance models are most appropriate for supervisory, predictive control applications necessary to achieve low-lift heat pump or chiller operation over a 24 h look-ahead schedule (Armstrong et al. 2009a, 2009b).

Armstrong et al. (2009a) developed a set of physics-based models for a variable-speed compressor-chiller in a low-lift cooling application. These consist of component models of a variable-speed reciprocating compressor, condenser and evaporator heat exchangers, an expansion valve, and a radiant-cooling distribution system, valid down to low-pressure ratios. The models can be used to predict chiller power consumption and COP for a given outdoor air temperature, zone air temperature, and cooling rate. Zakula (2010) built on this component modelling approach to include heat exchanger pressure drops and a static optimisation of condenser fan and compressor speeds, validated using data presented in this paper. Rather than use physics-based models, this paper focuses on empirical polynomial curve-fit modelling of heat pump and chiller performance suitable for use in a supervisory predictive control algorithm.

Empirical curve-fit models have been used extensively to represent steady-state heat pump and chiller performance. Stoecker and Jones (1982) presented an empirical bi-cubic curve-fit model of compressor power consumption as a function of refrigerant condensing temperature and evaporating temperature. This basic approach has been extended to create empirical multi-variable polynomial models of heat pump and chiller power consumption as a function of condenser fluid inlet (or outlet) temperature and evaporator fluid inlet (or outlet) temperature, as well as the cooling rate or part-load ratio, the ratio of cooling rate to reference cooling capacity under a given set of conditions.

The DOE-2 chiller model takes this approach to model chiller efficiency defined as electric input ratio (EIR), the reciprocal of COP, as a product of polynomials in chilled-water supply temperature, condenser water supply (or outdoor air) temperature, and cooling rate. The end result is a function that predicts power consumption as a product of two bi-quadratic polynomials in chilled-water supply and condenser-water supply temperatures and one quadratic polynomial in part-load ratio (Hydeman and Gillespie 2002).

Due to the increasing prevalence of chillers and heat pumps with variable-speed compressors and variable-speed condenser and evaporator flow, researchers have begun to adapt multi-variable curve-fit models to these more efficient variable-capacity systems. Hydeman and Gillespie (2002) noted that the existing DOE-2 chiller models have high errors in power prediction, around 10%, for variable-capacity chillers, particularly at low loads and low condenser temperatures—i.e., low-lift conditions. They also note that the DOE-2 chiller model is inadequate for systems with variable-condenser flow. Hydeman et al. (2002) present a modified DOE-2 chiller model in the same format as the DOE-2 model but with condenser-water return temperature as a variable instead of supply temperature. They also include temperature-dependent terms as well as third-order terms in part-load ratio in the polynomial modifying chiller performance for part-load operation. They found that this modified DOE-2 chiller better represented centrifugal chillers with variable-speed drives or variable condenser flow but did not present data on chillers with both variable-speed compressors and condenser flow.

Another approach to empirical curve-fit modelling of variable-capacity heat pumps and chillers is to apply multivariable polynomial functions directly to controlled variables, such as compressor speed or condenser fan speed. Shao et al. (2004) presented a curve-fit model of a variable-speed compressor that predicts power consumption and refrigerant mass flow rate as a product of a typical bi-quadratic in evaporator and condensing temperature and a quadratic in compressor speed with a correction for actual suction temperature. Application of this model to three compressors showed the ability to predict compressor power, mass flow rate, and COP to within 4% or less average relative error for all three compressors. The same approach was applied by Aprea and Renno (2009) to predict cooling capacity for variable-speed reciprocating compressors. However, these approaches were not extended to model more than just compressor performance (i.e., a full heat pump or chiller).

This paper presents an empirical curve-fit model for power consumption, cooling capacity, and EIR of a variable-capacity rolling-piston compressor heat pump as a quad-cubic function of compressor speed, condenser fan speed, outdoor air temperature, and zone air temperature valid at low-pressure ratios. This model is useful in supervisory predictive control applications, where compressor and condenser fan speeds can be optimised to achieve low-lift heat pump operation with radiant cooling and thermal storage. A detailed data set on the performance of a variable-capacity heat pump was needed to develop and validate this model.

The remainder of this paper will occupy three sections. The first section will describe the experimental test stand, test procedures, and validation of data on the performance of a rolling-piston compressor heat pump. The second section will describe quad-cubic polynomial curve-fits identified through linear regression on the experimental data to represent heat pump power consumption, cooling capacity, and EIR. The final summary and discussion section will review the findings of this research and describe applications for model-based predictive pre-cooling of thermal storage systems with variable-capacity chillers or heat pumps.

Experimental Measurement of Variable Capacity Heat Pump Performance

An experimental test stand was created using a rolling-piston compressor outdoor-unit heat pump and a finned-tube indoor-unit evaporator. The outdoor unit contains a rolling piston compressor with an accumulator, a finned-tube single-row condenser heat exchanger with a variable-speed condenser fan, and an electronic expansion valve. The indoor unit contains a finned-tube, double-row evaporator heat exchanger with a variable-speed evaporator fan. The working refrigerant is R410A. A schematic of the system is shown, and a picture of the system is included. Only cooling performance was tested.

The total thermal conductance across the zone control volume was measured to account for heat gains into the insulated enclosure, which contribute to the cooling load in addition to the electrical heaters. To measure thermal conductance, the temperature difference between ambient conditions (or, equivalently, condenser air conditions) and the air inside the zone control volume was measured using thermocouples installed inside and outside of the insulated box. A constant power was delivered to the heaters and fans inside the control volume, providing a constant heat rate. After a day of heating, a steady-state temperature difference was observed, from which the total thermal conductance could be calculated. Repeated tests showed the conductance of the zone control volume, [UA.sub.z], was approximately 1.9 W/K (3.6 Btu/h*F).

The volumetric airflow rate through the condenser was measured by traversing the condenser outlet air stream with a thermal anemometre. Samples of air velocity were taken at ten points along six radii following the methods for flow measurement outlined in ASHRAE Handbook of Fundamentals, chapter 14 (ASHRAE 2005a). Correlations between fan speed, airflow rate, and fan power consumption were established in order to relate airflow rate to fan speed. The ambient air pressure during each test was recorded from a local weather station, code KMACAMBR9, and along with temperature, was used to calculate the density and specific heat of the condenser air.

Steady-state performance data were collected at 131 chiller operating states spanning pressure ratios from 1.2 to 4.8, based on combinations of the conditions. The evaporator fan speed was fixed at the maximum speed. More generic testing could include varying the evaporator fan speed; however, evaporator fan speed was not varied for this research because the primary interest was the outdoor unit performance independent of the evaporator.

This view of the data shows the higher efficiency of the outdoor unit at low-pressure ratios. At typical pressure ratios, above about 1.8 or 2, [COP.sub.s] range from three to four, but at lower-pressure ratios the outdoor unit COP increases significantly to four to ten. The gains in the efficiency of the vapour compression cycle are somewhat offset by increased inverter losses at low speeds and increased fan power consumption. However, electrical inverter technologies have improved such that efficiency is high, in the range of 90 to 95%, over a broad range of operation. The compressor inverter efficiency as a function of pressure ratio (with compressor speed noted) on the left and the condenser fan inverter efficiency as a function of airflow rate on the right. The high efficiency of today's electrical inverters (or converters) over a broad range of speeds is an important enabling technology for high-efficiency variable-capacity heat pumps and chillers.

Energy and mass balances were performed on the heat pump performance data to validate the measurements at each steady-state condition. These calculations use all of the measurements shown except [w.sub.unit], [w.sub.(DC), (fan)], and [W.sub.(DC), (cmp*)] In a simple heat pump with zero piping and compressor jacket losses, the heat rejected at the condenser will equal the total heat absorbed at the evaporator, plus the total work performed by the compressor on the refrigerant or electrical power delivered to the compressor. The test stand includes measurements of condenser air volumetric flow, condenser air temperature difference and air pressure, from which to evaluate the specific heat and density of the ambient air in the condenser. It also includes measurements of the three-phase power delivered to the compressor and the cooling load on the evaporator. The cooling load on the evaporator includes both the electrical heater and evaporator fan power and the heat transfer into the zone control volume from the surrounding ambient conditions. With the foregoing measurements, a conservation of energy check can be applied to the three important heat and work transfers:

Condenser heat load: [Q.sub.condenser] = [rho]([P.sub.ao], [T.sub.aci]) [c.sub.p] ([p.sub.ao], [T.sub.aci]) [[for all].sub.ac] [DELTA][T.sub.ac] (2)

Evaporator heat load: [Q.sub.evaporator] = [W.sub.z] + [UA.sub.z]([T.sub.ao] - [T.sub.az]) (3)

Compressor power: [W.sub.(3[PHI]), (c)] (4)

In Equations 2-4, [rho]([P.sub.(ao), [T.sub.ac] and [c.sub.p]([P.sub.ac], [T.sub.ac]) are the pressure- and temperature-dependent ambient air density and specific heat.

The energy balance shown shows good agreement between the condenser heat load and the evaporator heat load plus compressor power. The relative root mean square error (RMSE), or the error divided by the range of variation, was 4.6% across all measurements; the relatively higher errors are present particularly at low loads. The three-phase power delivered to the compressor was used rather than total unit power or DC compressor power because the heat dissipated by the electronics, the inverters, and the condenser fan does not interact with the vapour compression cycle. The compressor power was measured using a high-frequency sampling digital power metre that accurately measures typical pulse-width modulated waveforms created by the inverters. The compressor and piping, apart from heat exchanger surfaces, were well insulated to minimise unmeasured heat transfers to or from the system.

A mass balance was also performed to validate the pressure and temperature measurements within the vapour compression cycle. The system is a single closed-loop system, and hence refrigerant mass flow rates must be the same through each component. Refrigerant mass flow rates through the compressor, condenser, and evaporator were calculated from measurements using the following equations:

Compressor mass flow rate: [m".sub.compressor] = [W.sub.(3[PHI]), (c)]/h([P.sub.rd], [T.sub.rd]) - h([P.sub.rs], [T.sub.rs]) (5)

Condenser mass flow rate: [m".sub.condenser] = [Q.sub.condenser]/ h([P.sub.rd], [T.sub.rd]) - h([P.sub.rs], [T.sub.rs]) (6)

Evaporator mass flow rate: [m".sub.evaporator] = [Q.sub. evaporator]/h([P.sub.rd], [T.sub.rd]) - h([P.sub.rs], [T.sub.rs]) (7)

In Equations 5-7, h is the pressure- and temperature-dependent refrigerant enthalpy for a given condition evaluated by REFPROP (NIST 2009) for R410A. The temperatures and pressures from which these enthalpies are calculate.

Two assumptions are implicit in the mass flow rate calculations. First, it is assumed that the discharge pressure is sufficient to calculate the enthalpy of the condensed liquid refrigerant exiting the condenser as an incompressible fluid. This is reasonable because in liquid state the refrigerant enthalpy is nearly independent of pressure. Second, it is assumed that the process of expansion through the electronic expansion valve is isenthalpic (adiabatic), and thus the enthalpy of the liquid refrigerant exiting the condenser is the same as the enthalpy of the two-phase refrigerant entering the evaporator.

A comparison of these three calculated mass flow rates to the mean flow rate is shown. The relative RMSE for the mass flow rates is 5.1% with large errors at some low-mass flow rate conditions. The compressor mass flow rate estimate was in error by as much as 19.5% under certain conditions. This may be a result of heat losses or mechanical inefficiencies in the compressor that could distort the measurement of discharge enthalpy, which is used to calculate mass flow rate. Another complication is the oil mass fraction and circulation rate. The oil mass fraction is difficult to measure and was not measured on the test stand. As a result, the enthalpies calculated from pressure and temperature measurements, which assume that the fluid is refrigerant R410A, may be in error due to the presence of oil mixed with the refrigerant (Willingham 2009). Equations 5-7 assume that the oil mass fraction is small enough that the change in refrigerant enthalpy completely dominates.

Overall, the test stand data show reasonable agreement in energy balance and mass flow rates. These data have been used by Zakula (2010) to develop improved physics-based heat pump component models for simulating low-lift compressor and chiller performance. The data will be used in the following section to develop empirical polynomial curve-fit models of heat pump performance.

Empirical Modelling of Variable-capacity Heat Pump Performance

As discussed in the literature review, multi-variable polynomial curve-fits long have been used to represent the performance of heat pumps and chillers. Recently, these empirical models have been adapted to variable-capacity chillers and heat pumps with inverter-driven variable-speed compressors and variable-speed condenser flow. A four-variable cubic polynomial was elected for modelling EIR, power consumption, and cooling capacity of the heat pump outdoor unit as a function of outdoor air temperature, zone air temperature, compressor speed, and condenser fan speed.

One must be careful when extrapolating polynomial models derived from experimental data outside the range of measured data. The data described above span a wide range of condenser air inlet temperatures and evaporator air inlet temperatures and a full range of compressor and condenser fan speeds. However, the full range of condenser fan speeds was not exercised at every possible combination of condenser air inlet temperature, evaporator air inlet temperature, and compressor speed. This is the result of a deliberate choice to limit the number of test points due to the time required for each test; however, the full range of the condenser fan speeds at each tested compressor speed was exercised for four different combinations of outdoor air temperature and zone air temperature.

To account for this limited number of tested condenser fan speeds, a deliberate quadratic dependence for the condenser fan speed variable was chosen. This dependence was observed for the pressure ratios and compressor speeds at which condenser fan speed was varied between 300 and 1200 rpm, and a similar dependence was assumed at other pressure ratios and compressor speeds. This quadratic dependence was enforced by eliminating high-order cross terms in the fan variable, such as cubic terms in the condenser fan speed variable. The minima of these quadratics can vary with compressor speed, outdoor air temperature, and indoor air temperature. The resulting curve-fit models have the following form:

In Equation 8, the dependent variable, DV, can be either the cooling capacity Q, the whole unit power consumption P, or the EIR = 1/COP at a given outdoor air temperature [T.sub.ao], indoor zone air temperature [T.sub.az], compressor speed [[omega].sub.c], and condenser fan speed [Florin]. The minima of the EIR across the condenser fan speed variable can be found by taking a partial derivative with respect to the fan speed. The optimal condenser fan speed as a function of outdoor air temperature, indoor air temperature, and compressor speed is given by the solution to Equation 9, using the coefficients for the 1/COP curve.

The resulting accuracies for this model choice are shown over the full range of data tested, which spans a wide range of pressure ratios, outdoor temperatures, and indoor temperatures as described above. These graphs show that a multi-variable curve-fit model can approximate the measured data well, with relative RMSE of 1.7%, 5.5%, and 4.7% for cooling capacity, power consumption, and EIR, respectively. The most inaccurate model is the power consumption model, which has an absolute RMSE of 27 W. At low compressor speeds, low-pressure ratios, and low power consumption, this can lead to inaccuracy as high as 10%-15% when the unit is consuming around 200 W. It may be possible to derive more accurate curve-fit models or look up tables for cooling capacity, power consumption, and EIR (or COP) as a function of the independent and controlled variables from physics-based modelling, for example (Zakula 2010).

The EIRs predicted from the curve-fit models are shown. The graph on the left shows the EIR of the air conditioner outdoor unit as a function of compressor speed at a fixed condenser fan speed of 700 rpm. Within each graph are multiple curves representing the EIR at a given combination of indoor and outdoor temperatures, [T.sub.az] and [T.sub.ao]. For a given condenser fan and compressor speed, and with fixed evaporator fan speed, [T.sub.az] and [T.sub.ao] are surrogate parameters to the evaporating and condensing temperatures in the vapour compression cycle. The graph on the right shows the EIR of the outdoor unit as a function of condenser fan speed at a fixed compressor speed of 50 Hz.

These performance curves can be used to predict the power consumption, cooling rate, and COP as a function of forecast outdoor temperatures and desired internal zone temperatures for given compressor and condenser fan speeds. Consequently, these curves can be incorporated into a predictive control algorithm in which compressor and condenser fan speeds are optimised to minimise energy consumption (or operating cost) subject to forecast outdoor conditions and desired zone temperatures. For buildings with significant thermal mass, pre-cooling can be performed to achieve more optimal low-lift operation at night as a result of this predictive compressor and condenser fan speed control. This will be explained in future papers.

Summary and Discussion

In this paper, an empirical curve-fit model of a vapour compression heat pump valid over a wide range of pressure ratios, from 1.2 to 4.8, has been developed from detailed steady-state performance data of a rolling-piston compressor heat pump. Existing empirical models for heat pumps and chillers, such as the DOE-2 model, are inaccurate at low-pressure ratios and for systems with variable-speed condenser fans or compressors (Hydeman and Gillespie 2002). New curve-fit models were identified that are valid at low-pressure ratios with a variable-speed compressor and condenser fan. Furthermore, these models explicitly include compressor speed and condenser fan speed as independent variables, making them well suited for use in predictive control algorithms that optimise compressor and condenser fan speeds.

The curve-fit models are four-variable cubic polynomial functions for cooling capacity, power consumption, and EIR as a function of outdoor air temperature, zone air temperature, compressor speed, and condenser fan speed.

Quadratic dependence on condenser fan speed has been enforced by eliminating third-order terms in condenser fan speed. The relative accuracies (in RMSE) of these models across the wide range of steady-state data are 1.7%, 5.5%, and 4.7% for cooling capacity, power consumption, and EIR, respectively, for the rolling-piston compressor heat pump that was tested.

Creating similar quad-cubic models of other variable-capacity heat pumps and chillers suitable for control applications can be accomplished in two ways. Physics-based models can be created that can be parameterized to accommodate different types and configurations of heat pumps and chillers (Zakula 2010). From these physics-based models, simplified quad-cubic curve-fits can be identified suitable for integration into predictive control algorithms. Alternatively, manufacturers may provide more detailed performance data for their products.

Currently, manufacturers provide data primarily at test conditions established by AHRI standards. Manufacturers could instead provide detailed data on the capacity and COP of heat pumps and chillers over a full range of compressor speed, condenser fan speed, evaporator fan speed, and indoor and outdoor air temperatures. This would enable advanced heat pump and chiller control strategies such as predictive pre-cooling for load curtailment and efficient low-lift chiller operation.

The models presented here have been developed to support predictive control algorithms that optimise heat pump or chiller compressor and condenser fan speed over a 24 h look-ahead period to efficiently meet cooling loads by pre-cooling thermal energy storage, such as through a concrete-core radiant floor (Gayeski 2010). The framework for this predictive pre-cooling control has been established in prior research (Armstrong et al. 2009a, 2009b), and results from an experimental implementation of this pre-cooling control will be described in future papers.

MacroAir Introduces New LEED-Compliant HVLS Fan for Home Use at Greenbuild Expo

MacroAir Technologies, Inc., the original inventors of HVLS (high volume low speed) fan technology, and the manufacturers of advanced, LEED-compliant six-blade HVLS commercial ceiling fans, today introduced its newest fan designed for use in the home. Two six-foot diametre Airvolution HVLS fans for residential use will be on display at the company's Greenbuild International Conference and Expo, booth number 3928 S. One fan will be hanging from the booth ceiling and running; the other will be on an open pedestal display that attendees can touch and look at from all angles.

The Airvolution HVLS fan for residential use is built around MacroAir's Direct Drive Technology that features a gearless drive and an electric motor. The technology overcomes a major design hurdle as it eliminates the gearbox, which is a point of failure that reduces fan life due to metal-to-metal contact. The only mechanical contact in the new Direct Drive are the bearings, which MacroAir designed and developed in-house specifically for use with the motor. The new Direct Drive Technology extends the expected life of the fan, improves energy efficiency, and offers silent operation. Fan diametres range from six-feet to 16-feet. MacroAir will begin shipping the new fans in early 2012.

"Our Direct Drive Technology is based on the drive motors used in electric vehicles," says Eddie Boyd, President of MacroAir. "It's advanced technology and it's expensive. However, we want to give our customers the benefits of using the latest technology - technology that's energy efficient and good for the environment. The motor on this fan now turns the blades, rather than the gearbox, which makes fan movement much more efficient. It also eliminates the heat generated by traditional gearboxes. This is, unquestionably, the most advanced HVLS fan on the market today and we're honoured to introduce it at Greenbuild."

Attendees at Greenbuild's International Conference and Expo will be able to view a moving Airvolution HVLS fan for residential use at MacroAir's booth 3928 S and an Airvolution 14-foot commercial HVLS fan. An Airvolution residential fan will also be mounted on a pedestal, allowing attendees to view the electric motor and airfoils. In addition to offering the Airvolution line of HVLS fans, MacroAir is the first HVLS fan manufacturer to develop and offer a 100% HVLS solar fan. MacroAir representatives will be on hand to answer questions.

"We built a brand new 20' x 30' booth to accommodate hanging HVLS fans," says Boyd. "For the first time at a tradeshow, people will be able to see a moving MacroAir HVLS fan - and experience for themselves how its slow movement and size provide a large column of air that efficiently cools down large spaces without the turbulence associated with small, high-speed fans."

Typically, residential fans are 57-inches in diametre and feature three to five blades. Because of their shape and size, they "chop" the air in much the same way a blender chops food by forcefully pushing small amounts of air outward. HVLS fans differ in both size and shape. Due to their unique airfoils, MacroAir HVLS fans move a large air mass very slowly (at approximately five MPH) down and outward in a uniform pattern, requiring far less energy and offering significantly more cooling benefits. One 24-foot HVLS fan can cool a 20,000 square foot warehouse. The HVAC industry is beginning to incorporate MacroAir HVLS fans into their cooling and heating systems as the fans help lower capital costs and allow building managers to raise AC set points while meeting LEED and ENERGY STAR requirements.

Fan Dance: Design is Emerging as a Major Factor in the Market

Consumers are paying increasing attention to the decor in every room of their homes and fans, as a result, are now seen as part of the furniture.

"There is a concern for both the esthetics and the functionality of fans," said Jeff Martin, group marketing manager for fans and dehumidifiers for Jarden Consumer Solutions. "Most of them are used on an end table in the living room, or on a dresser in the bedroom, small compact areas where you don't have a lot of space.

Therefore, looks—while not everything—are playing a larger role in shoppers' decisions on what fan to buy.

"In our experience, view and sense of the market, design becomes very important," said Martin Stadler, marketing director for Swizz Style. "People spend so much on interiors in general—interior design, furniture, kitchen, etc.—and now they are looking for household appliances of all types that match their home's style."

Design is one reason why this market has seen a significant shift in the types of fans that consumers are purchasing.

"Table fans have become a flat market," Martin said. "The share is being stolen from somewhere and I believe it's going to tower fans. These are more esthetic than table fans. They have a sleeker design for the living room, and you don't use them to move high volumes of air. Consumers are also more willing to spend a little more money for a better-looking product like a tower fan."

Stadler noted that one design element favouring tower fans is the fact that they don't take up much room. "Tower fans are very space saving, and this will make the future of the fan market as a whole," he said. "Of course, on design and materials there is still much to explore."

Design is not the only reason for the market-share surge from tower fans; another is technology. Stadler noted the recent introduction of bladeless fans, "which is basically a tower-fan technology using a round ring to blow out the air."

For the fan category in general, technological innovations have come few and far between. "There is a relatively slow pace of technology and innovation in table fans," said Cheri Wright, director of marketing for air purifiers at Kaz. "A bladeless fan is a good example of old technology applied in a new way. However, this technology was introduced at a substantially higher price point than most table fans, and limited distribution is expected." Wright added that bladeless fans entered the market last year at $300, versus an average fan's cost of $20.

The most popular operational innovations for fans are those that make them work more quietly. "Innovation that reduces sound level or improves power is generally more appealing to consumers," Wright said.

With the emphases on design and technology, the fan category remains price-sensitive with consumers. This can be seen by the fact that at retail, most consumers go to mass merchants and discounters to buy fans. "Over 65 percent of total table-fan units were sold in discount channels in 2009," Wright said. "The drug-and grocery-store channels are also major distribution points for table-fan sales."

Along with mass merchants, Martin identified hardware stores and home centres are two other key stopping points for consumer table-fan sales. He also described what he called the "consumer decision tree" for fan purchases: "First, the consumer defines a need. I'm hot, I need air circulation.' Then he picks a fan based on the use location.

Third, he decides where he's going to buy the fan. Then when he gets to retailer and looks over the products he makes a decision. He might be looking for table fans but see tower fans at the store, and that might change his mind. Then it comes to the price/value equation."

In Wright's opinion, price will be the prime motivator in fan purchases for the foreseeable future as well. "Going forward, the market will remain price-sensitive, she said. "Technology will continue to evolve at a slow pace, and basics will dominate."

The recovery of the economy will help the market, according to Martin. "If and when the economy recovers, the fan market will recover," he said. "It all comes down to consumers' comfort in spending money. They're choosing not to spend now because they don't know what the future holds.

Microchip Technology Expands Thermal Management Line with Proprietary Solutions for Fan Control

This fan-restart functionality enables the device to continuously initiate a restart during a fan-fault condition. This simplifies system design by eliminating the need for resetting the fan after a fan-fault condition occurs. The TC64xB devices also enable the user to set a minimum fan speed threshold (TC642B and TC647B) or an auto-shutdown threshold (TC646B, TC648B and TC649B) to control fan operation when the operating temperature falls to lower levels. In addition, the TC642/6/8B devices assert an over-temperature signal to indicate a possible over-heating condition.

Ideal for telecom and networking equipment, power supplies, office equipment, notebooks and servers, Microchip's TC64xB family of fan managers and fan-health monitors adjust fan speed according to temperature instead of continuously running a system's fan at maximum speed. Since the fan only runs as fast as necessary, the life of the fan is extended and the acoustic noise is reduced. Moreover, this family of devices detects when a fan locks up and launches an alert signal to the host processor. This enables the user to replace the fan in time or shut down the system, therefore eliminating any thermal issues that can impact the system's performance and reliability.

"Due to the increasing power dissipation of modern electronic systems and components, the implementation of fan cooling is becoming inevitable," said George Paparrizos, product marketing manager for the Analogue and Interface Products Division. "The TC64xB frees system processors from fan control duties, therefore saving costs and maximising system efficiency and reliability. This family provides solutions dedicated to control fan speed according to temperature."

Integrating FanSense(TM) technology, the devices enable designers to use 2-wire fans, therefore eliminating the need to purchase more expensive 3-wire fans. The FanSense technology is Microchip's patented design implementation of sensing when fan current pulses are present, thus detecting fan-fault conditions and protecting against fan failure, while aiding in system testing.

The TC64xB family utilises an external power device enabling design flexibility and allowing the devices to drive 5-, 12-, 24- and 48-volt fans rated up to 1 amp. Unlike linear regulation, the efficient pulse-width modulation (PWM) fan-drive leads to low-power dissipation.

These devices have an operating voltage range of 3.0V to 5.5V, an extended temperature range of -40(degree)C to +85(degree)C and are available in an 8-pin MSOP, SOIC and PDIP package. Available in quantities of 1,000-units, these devices are priced at $1.06 each for the TC647B, TC648B and TC649B and $1.12 each for the TC642B and the TC646B. Samples and volume quantities are available today.

About Microchip Technology

Microchip has integrated analogue technology, peripherals and features to meet today's demanding design requirements. Our broad spectrum of analogue products addresses thermal management, power management, mixed-signal, linear and interface solutions. Combined with "Intelligent Analogue" microcontrollers, Microchip offers an extensive analogue portfolio for thousands of high-performance design applications in the automotive, communications (wireless), consumer, computing and industrial control markets.

Microchip Technology Inc. manufactures the popular PICmicro(R) field-programmableRISC microcontrollers, which serve 8- and 16-bit embedded control applications, and a broad spectrum of high-performance linear and mixed-signal, power management and thermal management devices. The company also offers complementary microperipheral products including interface devices; microID(R) RFID devices; serial EEPROMs; and the patented KEELOQ(R) security devices. This synergistic product portfolio targets thousands of applications and a growing demand for high-performance designs in the automotive, communications, computing, consumer and industrial control markets. Microchip is headquartered in Chandler, with design facilities in Mountain View, Calif., Bangalore, India, and Lausanne, Switzerland; semiconductor fabrication facilities in Tempe and Chandler, Ariz., Puyallup, Wash., and Gresham, Ore.; and assembly and test operations near Bangkok, Thailand.

Microchip employs approximately 3,300 people worldwide and has sales offices throughout Asia, Europe, Japan and the Americas.

Note: The Microchip name and logo, PIC, PICmicro, microID and KEELOQ are registered trademarks of Microchip Technology Inc. in the United States and other countries. SPI is a trademark of Motorola. All other trademarks are the property of their respective owners.

Review of Residential Ventilation Technologies

This paper reviews current and potential ventilation technologies for residential buildings with particular emphasis on North American climates and construction. The major technologies reviewed include a variety of mechanical systems, natural ventilation, and passive ventilation. Key parameters that are related to each system include operating costs, installation costs, ventilation rates, and heat recovery potential. The paper also examines related issues, such as infiltration, duct systems, filtration options, noise, and construction issues. This report describes a wide variety of systems currently on the market that can be used to meet ASHRAE Standard 62.2-2004, Ventilation and Acceptable Indoor Air Quality in Low-Rise Residential Buildings. While these systems generally fall into the categories of supply, exhaust, or balanced, the specifics of each system are driven by concerns that extend beyond those in the standard and are discussed. Some of these systems go beyond the current standard by providing additional features (such as air distribution or pressurisation control). The market will decide the immediate value of such features, but ASHRAE may wish to consider related modifications to the standard in the future.

Introduction

The purpose of ventilation is to provide fresh (or at least outdoor) air for comfort and to ensure healthy indoor air quality by diluting contaminants. Historically, people have ventilated buildings to provide source control for both combustion products and objectionable odors (Sherman 2004a). Currently, a wide range of ventilation technologies is available to provide ventilation in dwellings, including both mechanical systems and sustainable technologies. Most of the existing housing stock in the US uses infiltration combined with window opening to provide ventilation, sometimes resulting in overventilation with subsequent energy loss, sometimes resulting in underventilation and poor indoor air quality. Based on the work of Sherman and Dickerhoff (1998), Sherman and Matson (2002) have shown that recent residential construction has created tighter, energy-saving building envelopes that create a potential for underventilation. Infiltration rates in these new homes average three to four times less than rates in the existing housing stock. As a result, new homes often need ventilation systems provided to meet current ventilation standards. (McWilliams and Sherman [2005] and McKone and Sherman [2003] have reviewed such standards and related factors.)

According to ANSI/ASHRAE Standard 62.2-2004, Ventilation and Acceptable Indoor Air Quality in Low-Rise Residential Buildings, published by the American Society of Heating, Refrigerating and Air-Conditioning Engineers (ASHRAE 2004), single, detached residential buildings are required to meet a whole-house ventilation rate based on the number of bedrooms in the house, the number of occupants, plus an infiltration credit (3 cfm per 100 [ft.sup.2] plus 7.5 cfm per additional occupant, which includes a 2 cfm per 100 [ft.sup.2] allowance for infiltration). There are a variety of ways to meet this standard, either through mechanical systems or via natural forces. But for some occupants and homeowners, there is more to ventilation than just meeting the standard. They may desire added features for the comfort and health of the indoor environment or to reduce energy costs. According to Home Energy Magazine (Rudd and Lstiburek 2001), a good ventilation system should;

* provide a controlled amount of unpolluted outdoor air for both comfort and dilution,
* have at least a 15-year life,
* be acceptable to operate by occupants (low noise, low cost), and
* not detract from the safety and durability of the house.

This paper will review both mechanical and sustainable ventilation technologies and the factors that affect their effectiveness. Mechanical technologies include:

* continuous exhaust systems,

* intermittent exhaust systems,
* exhaust system with makeup air inlets,
* local exhaust with outside air integrated in HVAC system,
* continuous supply system,
* intermittent supply with inlet in return side of HVAC system,
* combined exhaust and supply (balanced) system, and
* houses without central forced-air distribution systems.

Sustainable technologies, which are those whose motive forces are principally temperature difference and wind, are reviewed later in this paper and include :

* infiltration with operable windows,
* passive stack ventilation,
* solar chimney, and
* hybrid systems.

The effects of incidental ventilation provided by infiltration and operable windows are discussed. Finally, a variety of factors that can affect ventilation effectiveness are discussed, including cost and energy use, air cleaning and filtration, construction quality, control systems, and duct systems.

Mechanical Whole-house Ventilation

There are a variety of mechanical whole-house ventilation systems, including exhaust, supply, and balanced systems. Any of these may be in continuous operation or operate intermittently, they may be single-port or multi-port, or the system may be integrated into an existing HVAC system. Mechanical ventilation strategies provide more uniform ventilation rates than natural ventilation (Hekmat et al. 1986). Properly designed mechanical systems provide good control over ventilation rates when compared to most other ventilation systems; however, additional energy is required to operate the system. Holton et al. (1997) compared ventilation systems in newly built homes and found infiltration rates ranging from 0.1 to 0.07 ach in the summer and 0.35 to 0.15 ach in the winter. As a result, they recommend that modern houses include a mechanical ventilation system. Researchers have studied various configurations of exhaust, supply, and balanced ventilation systems, with and without whole-house recirculation by the central heating and cooling air-handler fan, and these are reviewed in the following.

Continuous Exhaust System

A continuous whole-house exhaust system provides ventilation by using a single-point or multi-point central fan to remove air from the building (Concannon 2002). Supply air enters the building envelope through gaps or provided vents. If the building envelope is tight, there is a possibility that negative pressure can be created inside the building, leading to backdrafts from combustion (open flue) appliances. Often these systems incorporate a pressure relief damper to alleviate pressure imbalances. Supply air enters the building in an uncontrolled manner and may be pulled in from relatively undesirable areas, such as garages, musty basements (or crawlspaces), or dustyattics (Barley 2002). Whole-house exhaust systems may not be appropriate in areas where levels of outside environmental contaminants are high. In the case of radon, researchers have found that exhaust systems may actually increase the indoor levels of contaminants (Bonnefous et al. 1994). In severe climates, very cold supply air may create drafts, while in moist, humid climate zones, exhaust-only systems can cause moisture damage to the building structure. Filtration cannot be sensibly added to an exhaust-only ventilation system unless one considers the building envelope as part of the filtration system.

Heat recovery can be added to exhaust systems: Passively, the building envelope itself can provide some heat recovery (Walker and Sherman 2003b) and is also partially effective at removing ozone. More actively, an exhaust air heat pump can be used to recover the energy in the exhaust airstream.

The Home Ventilating Institute (HVI 2005) lists a large variety of fans that can meet current ASHRAE standards for ventilation rates if properly installed. However, several factors (such as the tightness of the building envelope, size, quality of ductwork, and placement of ducting, among others) can have a significant effect on whether the installed fan can provide the indicated ventilation rate. These fans can potentially provide ventilation rates from 50 cfm to more than 5000 cfm.

Most of the operating costs result from conditioning the supply air rather than from the energy to operate the fan. The HVI directory lists the energy use for only a small percentage of the fans, with typical power consumption of about 3.5 cfm/W. Wray et al. (2000) found that from most perspectives, exhaust-only mechanical ventilation systems are the most inexpensive mechanical systems to operate.

Single-Point Exhaust System: A single-point exhaust system is often an upgraded bathroom fan. Construction and installation costs are the lowest of the mechanical systems (Concannon 2002). Only one fan and possibly some simple ducting are required to exhaust the air to the outside.

In some cases, the fan can be installed in an exterior wall, eliminating the need for extensive ductwork. Single-point ventilation systems suffer from a nonuniform distribution of fresh air, especially to closed rooms (Rudd and Lstiburek 2000). In an evaluation of five mechanical ventilation systems, Reardon and Shaw (1997) found that local exhaust-only strategies (which depended on kitchen and bathroom fans to provide whole-house ventilation) provide only marginally better performance than infiltration alone. This simple system suffers from a poor distribution of supply air. Standard 62.2-2004, however, has no distribution requirement; so this is not an issue for a minimally compliant system, but it is nevertheless a consideration.

Multi-Point Exhaust System: Multi-point exhaust systems are an improvement over single-port exhaust systems in that they improve the room-to-room uniformity of the whole-house ventilation, but there is extra cost required for the installation of the ductwork (Rudd 1999). One exhaust fan is ducted to many rooms of the house and may be remotely installed to reduce noise levels. In a comparison of ventilation systems, Reardon and Shaw (1997) found that if a centralised multi-point system was installed, air was distributed evenly throughout the house, even to closed bedrooms.

Intermittent Exhaust System

An intermittent exhaust system is similar to a continuous exhaust system; generally it consists of one central fan to remove stale air from the building, but it may also incorporate several fans in areas of high sources (i.e., bathrooms and kitchens). In this case, the fan(s) runs only part of the time at a higher rate and is sized to provide the necessary ventilation.

The rate of ventilation when the system is operated intermittently must be larger than if it were operating continuously (Sherman 2004b). There are several advantages for using an intermittent ventilation system. The occupant can reduce the amount of outdoor air entering the building during periods of the day when the outdoor air quality is poor. Peak load concerns may make it advantageous to reduce ventilation for certain periods of the day. When the ventilation system is integrated with the heating and cooling system, cyclic operation may also make more sense.

The occupant can control the fan when needed: The disadvantage here is that the occupant controls the ventilation and must be relied on to know when ventilation is needed. The occupant may choose not to operate the system (for example, if the fan is noisy), which could result in underventilation. (If the system is Standard 62.2-2004-compliant, ventilation fans should meet sound requirements and noise should not be a substantial issue.) Many systems use a timer to automatically run the fan for a certain amount of time each day so that the occupant is not relied on to sense when ventilation is needed. However, the occupant often has control over a switch to turn the fan on high when extra ventilation is needed. More sophisticated (and costly) control systems are available, including C[O.sub.2] sensors, occupant sensors, and humidity sensors. C[O.sub.2] sensors and occupant-controlled systems do not meet the current Standard 62.2-2004 requirements unless those features are used to raise the ventilation over and above the minimum rates required by the standard.

Our own experience has shown that installation and operating costs are similar to the continuous exhaust systems but may exceed them if sophisticated control systems are installed. As with continuous exhaust systems, most of the energy requirements are for conditioning the supply air rather than fan operation. The potential exists to reduce energy consumption when compared to the continuous exhaust system if the intermittent system is used in conjunction with natural driving forces to provide adequate ventilation while reducing the energy required to condition outside air. For example, running the fan at night could reduce cooling costs. Also, the fan could be programmed to run during times when outside pollutant levels are low or, alternatively, to shut down the system when outside particulate or ozone levels are high. If time-of-use utility rates are locally in use, it may be possible to reduce operating costs by ventilating more during low-cost periods to allow reduced or even zero ventilation during high-cost periods.

Exhaust System with Makeup Air Inlets

Another mechanical ventilation system uses an exhaust fan but controls the entry of supply air into the dwelling by providing openings specifically for air supply. Trickle vents, air inlets, or louvers can be located in rooms that need extra ventilation, such as the bathroom. Again, filtration of the supply air is not possible with this system; however, the entry point of the supply air can be controlled to provide cleaner air by installing trickle vents away from polluted areas such as garages, musty basements, or dusty attics. Trickle vents are not necessary to meet Standard 62.2-2004 per se but may be needed in exceptionally tight construction to reduce depressurisation and related issues. They are commonly used as part of European systems both because of the tight construction and to ensure that habitable rooms have individual air supplies.

Local Exhaust with Outside Air Integrated in HVAC System

This method builds on the exhaust systems described above but adds an outside air inlet in the return duct system of the air-handling unit of the HVAC system. Depending on the path(s) of least airflow resistance, this may use the existing duct system to extract air from individual areas of the building. Because existing ductwork would be used, marginal installation costs can be kept very low. This system can provide uniform ventilation throughout the house and may be operated intermittently or continuously. There are added operating expenses for running the central fan when heating or cooling is not needed, which depends on climate and system sizing.

Often exhaust systems are designed to comply with Standard 62.2-2004, unlike the centrally integrated makeup air system that can provide not only ventilation but also air distribution and can counteract depressurisation—both of these later attributes go beyond the minimum requirements of Standard 62.2-2004 and are often desirable. In principle, the makeup air system could be designed to meet Standard 62.2-2004, and the exhaust system could be used as a source control enhancement.

Continuous Supply System

Continuous supply systems allow the occupant to control the location of the supply air to maximise air quality and give the occupant the option of filtering and/or conditioning the supply air (Building Science Corporation). Air is supplied by a central fan ducted to some or all of the rooms of the dwelling, forcing stale air out through leaks in the building envelope. This system creates a positive pressure inside the building, which has both advantages and disadvantages. The amount of the pressure depends on the supply flow and the tightness of the envelope. A positive pressure prevents outside contaminants from entering the building, but it also can force moisture-laden air through the building fabric. In cold climates, the moist air may condense in the walls of the building, creating an environment for mold growth. Various studies have considered the use of whole-house fans to provide night ventilation for cooling purposes (Santamouris 2006). In these systems, air-conditioning loads may be reduced up to 56% depending on the thermal preferences of the occupants.

Because outdoor air is often not in the thermal comfort zone, the temperature of the supply air is a design concern. Supply systems need to address this concern by conditioning or tempering the air in some way during the periods when it would be perceived as unacceptable. One method, for example, is to mix the supply air with indoor air before it reaches the occupants. Standard 62.2-2004 has no requirements for tempering. As with continuous exhaust systems, there are two main designs: single-point and multi-point systems.

Single-Point Supply System: In this strategy a supply fan provides fresh air via a small amount of ducting to a main room of the house. The air is distributed about the house by natural process. Often there is a return duct in a separate room. This system has low equipment costs; only the fan and a small amount of ducting are needed. However, the system suffers from a poor distribution of supply air, especially to closed rooms in the house (Rudd and Lstiburek 2000) even compared to single-point exhaust. Tempering or conditioning of this air is almost always needed if one wishes to avoid comfort complaints.

Multi-Point Supply System: The multi-point system has the advantage of improving ventilation uniformity throughout the house but with the extra installation cost of the ductwork. Because each supply is of a lower flow, the needs for tempering or conditioning may be reduced. From the perspective of Standard 62.2-2004, however, there are no differences between single- and multi-point supply systems.

Intermittent Supply with Inlet in Return Side of HVAC System

Integrating the supply air into the existing HVAC system provides a low-cost option to supply and distribute fresh air through the existing duct system and is the ventilation system most acceptable to large production home builders (Rudd and Lstiburek 2001). In this system, the existing central forced-air system is used to supply fresh air in a distributed manner through the building's ducting. An outside air inlet is placed in the return of the HVAC system to allow fresh air to enter when the air handler fan operates.

All mechanical ventilation systems benefited from intermittent operation of the central fan. This resulted in more uniformity of ventilation air among the various rooms of the house (Rudd and Lstiburek 2000). By operating the ventilation system intermittently as opposed to continuously, Rudd (1999) estimated a 28% annual savings in total energy use. Computer modelling studies showed the cost-effectiveness of this system when compared to a separate supply ventilation system as well as the marginal costs of operation compared to no mechanical ventilation ($3 to $27 per year) (Rudd and Lstiburek 1998). Plus, they estimated it would take ten years to recover the initial costs of a separately ducted supply ventilation system. According to the computer modelling, the continuous and intermittent simulated systems had average outside air exchange rates of between 40 and 50 cfm, including the combined effects of ventilation and infiltration. These rates met Standard 62-1989 but would not meet the current Standard 62.2-2004.

A supply system can create positive pressure in the house, so a pressure relief vent is often installed. We often see pressure relief achieved through backdraft dampers of bathroom and kitchen exhaust fan ducting, as well as incidental leakage sites around windows, doors, or other building penetrations.

It is possible to add filtration to the supply air to remove contaminants. Installation costs are minimal for the return inlet itself; only a small amount of extra ducting and possibly a damper are required. Depending on the design, extra costs may be incurred for control devices and/or dampers.

Heat recovery potential for intermittent supply is low since heat exchange only occurs as the exhaust air exits via exfiltration through the building fabric. Currently air-handler fans are available to meet airflow rate standards in an energy-efficient manner. Simple control systems are available to operate the system when the HVAC system is heating or cooling or to operate the system on a timer so that fresh air is supplied when heating or cooling are not required (Walker and Sherman 2003a, 2003b). Energy efficiency is maximised when the entire air distribution system is airtight and located in a conditioned space (Rudd and Lstiburek 1998).

Combined Exhaust and Supply (Balanced) Systems

A balanced ventilation system uses two fans with separate ducting systems, one to supply fresh air and one to remove stale air from the building. The system should not affect the pressure balance of the interior space unless the return path between the supply and exhaust is blocked. This ventilation strategy can be used effectively in any climate. It is possible to include a heat exchanger (or heat pump) to recover heat from the exhaust air and use it to precondition the supply air. Extensive ducting is used to supply fresh air to living and sleeping rooms, while a separate exhaust system removes stale, often moist, air from the kitchen and bathrooms. Advantages include prefiltration of the supply air and energy savings from the heat recovery of the exhaust air. Some disadvantages include installation costs, maintenance costs (because there are multiple fans), and possibly fan noise (for fans not meeting Standard 62.2-2004 noise requirements). Noise generated from the fan(s) and ducting system can be transmitted to each room of the house and reach 30 to 40 dB. Veld and Passlack-Zwaans (1998) describe various strategies for soundproofing, including insulating ducts and preventing fan vibrations. Reducing noise from ventilation systems has a positive impact on indoor air quality by reducing the likelihood that occupants will block vents or turn off the system.

Supply Integrated into HVAC System with Continuous Exhaust. If the house has an existing central forced-air system, it is possible to save on installation costs by integrating the supply inlet into the return of the HVAC system. A separate exhaust fan would run continuously to remove the stale air. This system can sometimes be problematic in humid climates, where moist air is injected into cool supply air ducts, resulting in condensation, and independent humidity control may be required.

Supply Integrated into HVAC System with Intermittent Exhaust. In this strategy (similar to the above) the exhaust fan would operate intermittently. Advanced control strategies can in principle be used to operate the exhaust fan only as needed to supplement the supply air to the return.

Houses without Central Forced-Air Distribution Systems

Most new homes in the US are built with central forced-air systems, but not all. Houses with radiant, hydronic, and/ or baseboard systems may not have any central air distribution system and cannot use any of the HVAC-integrated systems discussed above. Any of the other systems, however, can be used to meet Standard 62.2-2004.

If air distribution is a concern, however, some systems may perform better than others for houses without central forced-air systems. If the building envelope is tight, an exhaust system with trickle vents or air inlets can increase the likelihood that each room will get some outdoor air. Supply or balanced approaches require a dedicated distribution system (i.e., multi-point supply) in order to achieve good air distribution.

Sustainable Ventilation

All of the systems described above focus on a mechanical ventilation solution. Standard 62.2-2004 does not mention any other way to provide ventilation to new construction, but it does allow (in Section 4.1.2) alternative approaches if approved by a licensed design professional. There is a variety of potential ventilation options that do not require fans. Here we examine such sustainable technologies with the understanding that they do not meet Standard 62.2-2004, but they do allow advanced solutions in the future.

Tradition: Infiltration with Operable Windows

Many existing homes rely on infiltration through a porous building envelope for background ventilation with operable windows to provide increased ventilation when needed. Natural climatic forces create differences in air pressure between the outside and inside of the building, which can ventilate a building. Pressure differences depend on changes in temperature and wind speed. Wind causes a positive pressure on the windward side of the building and a negative pressure on the leeward side of the building. The resulting amount of ventilation is dependent on the placement and number of openings in the building envelope and on wind direction and speed. This makes the ventilation rate unpredictable and uncontrollable since the driving mechanism is variable over the year and the flow paths are diffused over the building envelope (Allard and Ghiaus 2006). The average ventilation rate may be predictable, but the average ventilation rate itself is not the key factor.

Sherman and Matson (1997) have shown that typical existing homes have an annual average air change rate of over one air change an hour due to infiltration; this high rate can satisfy existing ventilation standards so that many existing homes do not need an extra ventilation system. Dwellings in cold, harsher climates and new residential construction are three to four times tighter, creating a tight building envelope and the potential for underventilation (Sherman and Matson 2002).

This basic system of operable windows has no extra construction costs or explicit operating costs. The energy implications are almost exclusively from the need to condition the outdoor air. The system relies on the occupants to open and close windows to provide adequate ventilation, particularly when the building envelope is tight; however, there is poor control over ventilation rates when the envelope is leaky. The lack of control can result in energy loss due to high air change rates, especially in winter when temperature differences and wind speeds are high. Alternatively, the system may underventilate during the hot summer months. When climatic conditions are favourable, natural ventilation can be used for cooling and can replace air-conditioning systems for part of the year.

But in urban settings there are considerable limitations to such an open ventilation system including noise, security, and pollution (Santamouris 2006). Additional limitations arise from the unique climatic conditions of cities. Both higher temperatures (the heat island effect) and decreased wind speeds in urban canyons can decrease the potential of natural ventilation systems. Geros et al. (2001) studied the reduction of airflow in naturally ventilated buildings in ten urban canyons in Athens, Greece, and found that because of the reduced wind speed, the airflow through the buildings decreased up to 90%. A few strategies exist for reducing noise in buildings using operable windows, and they are capable of reducing traffic noise by 7.5 to 8.5 dB without compromising the airflow path resistance (Oldham et al. 2004).

Since climate plays an important factor in the effectiveness of natural ventilation, many groups have analysed the suitability of various climatic conditions. The potential of natural ventilation depends not only on the outdoor climate but also the building site and the design of the building site. Yang et al. (2005) have created a model to evaluate the potential of a particular site to provide the natural forces necessary to meet ventilation standards with only natural ventilation. It is clear that many climates are too harsh for infiltration to be used as a primary source of ventilation. Conversely, there are climates where the driving forces are too weak for infiltration to be a practical source of primary ventilation. All of which leads Wilson and Walker (1992) to conclude, "There is no hole for all seasons."

Infiltration does provide ventilation automatically without using any transport energy, but it almost always requires more space-conditioning energy to supply the equivalent ventilation as a constant mechanical system. Infiltration can provide some heat recovery and filtration through the building envelope, but unless it is well designed (e.g., the "dynamic insulation" used in Scandinavia) it is not likely to provide very much. Infiltration depends on the weather, so there is no "right" amount of air leakage. Infiltration will always provide more ventilation than is needed during extreme periods in order to meet average demands.

Passive Stack Ventilation

Passive stack ventilation is designed to provide more control over natural ventilation rates by incorporating one or more stacks or towers into the building structure to extract stale air while fresh air enters through provided openings such as trickle vents or louvers. Passive stack airflows are created from a combination of two climatic forces: differences between the inside and outside temperature and wind speed. The negative pressure at the stack top is often the critical factor. As shown, wind speed creates negative pressure on the leeward side of the building in many situations. The combination of cooler, entering air with warmer, less dense indoor air and negative wind pressure at the top of the stack result, in air being exhausted from the stacks.

Although rare in the United States, passive ventilation systems are widely used in the European Union. Axley (2001) found that in England and the Netherlands most single-family dwellings use passive ventilation (90% and 65%, respectively). Emmerich and Dols (2003) have used some of Axley's approach to create a passive ventilation design and analysis tool for use in a multizone environment.

Stack height and position are important in maintaining a negative pressure at the stack terminus and preventing backflows into the building. A taller stack is less sensitive to wind speed and wind direction. Installation guidelines and building codes reflect the importance of stack position relative to the roof. A variety of terminal caps are available that are designed and located to provide consistently negative pressures (independent of wind direction) at the stack exit (Axley 2001). Stacks need to have a larger diametre than mechanical ducting systems to reduce flow resistance for low pressure drop conditions. At present there is insufficient information to recommend specific minimum or maximum values for performance parameters, but there are references worth considering, including those in the AIVC database and in Stephen et al. (1994). A draft European Standard for testing cowls and roof outlets is in preparation (prEN 131415). Ventilation flow rates can vary significantly from room to room. Upper, leeward rooms in particular may be underventilated and can easily have no outdoor air. Careful design measures can be taken to control and distribute flow rates. Typically, systems are designed with trickle vents or louvers that can be manually adjusted to control the flow rate, but these work best when uncontrolled infiltration rates are low (and building envelopes are tight). Each room must have a transfer grille or vent to allow free distribution of the air. While these same criteria are relevant for mechanical ventilation, the issue is often more critical for passive ventilation because of the lower driving forces. Many anecdotal cases indicate that passive ventilation systems have shown the capability of providing adequate long-term ventilation but fall short when required to provide short-term high ventilation during peak episodes of contaminant production (i.e., bathing or cooking).

Because they are designed similarly to mechanical systems but without mechanical components, passive stack ventilation systems can reduce construction and operating costs of residential buildings. Careful design of internal spaces should be considered during construction to allow air to flow between the rooms of the building and from the supply openings through the exhaust spaces. Relatively larger ducts than those used in mechanical systems are required since flow resistance is an issue. Operating (air transport) costs are nonexistent, however, there are usually some days of the year when weather conditions (low wind speed and/or small indoor/outdoor temperature differences) create insufficient airflow.

There is inherently some uncertainty in any system performance that is dependent on natural driving forces. Underventilation or overventilation can be expected at certain times of the year (Yoshino et al. 2003). Wilson and Walker (1992) showed that even with several large passive ventilation openings, single-family residences could not be adequately ventilated (relative to Standard 62-1989) during periods of light winds (less than 10 km/h) or small temperature differences ([DELTA]10[degrees]C). These conditions are common in the spring and fall. At these times proper ventilation may only be attained if the occupant opens a window or otherwise supplements the system. The usual natural forces are highest on cold days, creating overventilation, cold drafts, and energy loss. Self-regulating vents are available that can reduce or control overventilation. Pressure-sensitive ventilators are available that can provide constant ventilation rates over a wide range of pressures, but these passive control units are relatively scarce (Axley 2001). Passive systems fall short when compared to mechanical systems in the areas of filtration and heat recovery. Filtration of the supply air is not feasible, and heat recovery is also relatively uncommon. Shao et al. (1998) have shown that heat pipes can be used with 50% heat recovery efficiency. Another strategy for heat recovery is to install inlet vents into the subfloor. This strategy will temper cold supply air and help avoid cold drafts and will also reduce the sensitivity of the ventilation rate to wind direction (Hayashi and Yamada 1996).

Solar Chimney

A solar chimney is a passive stack system fitted with a solar collection panel (or often glazed walls on the south side of the building) that is used to heat the air in the stack, resulting in an increased buoyancy of the air in the stack. By increasing the temperature differential between the inside and outside of the stack, ventilation rates are substantially improved on warm, sunny days with low wind speeds (Bansal et al. 1994). This can improve the year-round effectiveness of the passive stack ventilation system. Airflow rates can be increased 20% over passive stacks without solar chimneys (Jaros and Charvat 2004). Khedari et al. (2003) reported that a solar chimney could reduce the load on the air-conditioning system (by using ventilative cooling), resulting in an average electrical savings of 10% to 20%. The advantages of this system are the increased reliability of the passive stack system plus the system being silent and transparent to the occupant. The disadvantages are the extra design, installation, and cost of the solar glazed panels. This system would be most appropriate for a sunny, warm climate.

Hybrid Systems

Hybrid systems are passive systems with a low-power fan to boost the flow of air through the stacks or vents, thus combining the advantages of a passive system with the reliability of a mechanical system. The combination of the two systems improves the indoor air quality while reducing energy demand through an intelligent controller (Heiselberg 2006; Li and Heiselberg 2003). There are a number of ways the two systems may be combined. The building may have two independent systems linked by a controller to switch from one to the other (a mechanical exhaust fan for the summer and winter and natural ventilation for the moderate seasons, for example). Another combination is fan-assisted natural ventilation, where the main ventilation is provided by natural forces but a low-power fan can be switched on to assist ventilation during periods of weak natural forces. A third, similar strategy is to include a small fan in a passive stack system to assist in creating optimal pressure differences in the stack. Yoshino et al. (2003) have shown that a hybrid system can provide adequate ventilation rates even when weather conditions create poor ventilation in the passive system. By using a fan to boost stack ventilation during times of low wind speed, underventilation was prevented. And by using damper control at the vents, overventilation was prevented when temperature differences were large. Often, these systems incorporate the use of sophisticated control systems such as CO2 sensors, room temperature, airflow sensors, motorized windows, and even a weather station (Dorer et al. 2004a). Filtration of supply air is not common. The main disadvantage of hybrid systems is the complex control system. This adds an extra cost to the installation in terms of expensive parts and trained personnel to install them. Most occupants feel comfortable with (or prefer) a simpler user interface.

Incidental Ventilation

Incidental (or adventitious) ventilation refers to features or effects that were not designed to provide whole-house ventilation but, in fact, may. When they are truly incidental, one does not "count" them in the ventilation design, but one may need to take account of them in order to determine the actual energy and indoor climate impacts of a specific design.

For example, an air-to-air heat exchanger can only recover the energy of the air that goes through it. If the building is leaky and a significant fraction of the actual ventilation air bypasses the exchange, energy performance will be severely compromised. By contrast, the performance of an exhaust air heat pump is less dependent on envelope airtightness, although not completely independent.

Infiltration

Air leakage through the building envelope can have a detrimental effect on ventilation effectiveness regardless of the ventilation system. Infiltration rates are not constant since they are dependent on the weather. Both mechanically and passively ventilated leaking homes will lose energy when infiltration rates are high during the heating season. Very little heat recovery occurs in the building envelope (Walker and Sherman 2003b), which generally results in a loss of energy used to condition the infiltrating air. Balanced ventilation systems will also suffer a reduction in performance when air bypasses the heat recovery unit. However, buildings that are too tight may also suffer from a reduction in indoor air quality. Mechanical exhaust systems can create a negative pressure inside the dwelling when infiltration is low. This can lead to back-drafts from combustion appliances, poor indoor air quality, and high fan power requirements.

There are several methods available to measure leakage of the building envelope (Sherman 1990; Sherman and Chan 2006; Ask 2003; Dorer et al. 2004b). Ideally a building would leak no more than the air required for healthy indoor air. The amount of infiltration will depend on the airtightness of the building, the difference in indoor and outdoor temperatures, and the wind pressure. A tight building envelope will provide very little ventilation from infiltration and will require a ventilation system. Infiltration rates need to be taken into consideration when designing an HVAC system.

Sherman (1995) created a map of infiltration zones required to meet Standard 62-1989 ventilation standards based on the climatic data of each zone. In mild climates (such as the coast of California) infiltration alone is not enough to provide adequate ventilation in newer well-insulated homes, while in harsher climates, infiltration rates may be so high as to cause overventilation, energy loss, and comfort issues due to drafts. This zone would require the tightest home construction.

Operable Windows

Most homes are required to have operable windows in each room of the house. Occupants are more likely to feel comfortable when they have control over the ventilation system, and windows provide a familiar system of ventilation. If used on a daily basis, windows can provide the ventilation necessary to meet current codes. Liddament (2001) reviewed several studies on occupant behaviour and ventilation and found that windows were most likely to be opened under the following conditions: sunny days, higher occupant density, higher outdoor temperature, low wind speed, during cleaning or cooking activities, and when smoking. However, there are many circumstances when opening a window is not practical, such as noise, rain or high winds, outdoor pollutants, cold drafts, privacy, security and safety issues, and energy loss, or the window may be difficult to operate. These observations suggest that window opening or closing is not always in response to ventilation needs.

Local Exhaust Fans

Local exhaust fans are often used in rooms with high moisture to provide source control when needed, most commonly kitchens and bathrooms, but laundries, utility rooms, and lavatories may also have local exhaust fans. Local exhausts fans are not intended to dilute contaminants but rather to remove them while they are still concentrated. As such, they are a source control measure rather than ventilation in the normal sense.

While doing their source-removal job, they may also increase the overall ventilation of the building and, in that sense, are incidental ventilation. For example, a high-capacity kitchen exhaust of 400 cfm ensures that the overall ventilation rate will be (temporarily) at least 400 cfm, which is well above minimum Standard 62.2-2004 rates. Because the duty cycle of these local exhaust fans is determined by the occupants and presumably related to a source-generating activity, one cannot count on them toward meeting minimum ventilation requirements.

A notable exception to that last statement is the "double duty" bath fan. In this design, a continuously operating local exhaust fan simultaneously meets the need for local exhaust and also whole-house ventilation. Provided the fan meets the appropriate requirements (e.g., sizing, noise), Standard 62.2-2004 allows this approach.

Real-world Factors

Standard 62.2-2004—or any other ventilation standard or code—is a set of minimum requirements that, if followed, will provide a certain minimum level of indoor air quality. In deciding how to apply such requirements, however, a variety of real-world factors need to be considered, such as construction, installation and energy costs, control and distribution systems, and indoor air quality. Often these decisions are determined by the needs of the client (or builder) more so than the requirements of the standard (Rudd and Lstiburek 2001).

Construction and Installation Issues

A potential problem exists when ventilation technologies are not properly installed or designed (Dorer and Breer 1998). Any ventilation system will not reach its performance potential if components are poorly manufactured or installed improperly. In 2001 a group of recently constructed homes in Minnesota were examined for various performance measures. Sheltersource, Inc. (2002) found that the average measured bathroom fan exhaust capacity was only 71% to 75% of the total rated capacity. Several factors contributed to poor performance, including long duct lengths. Compression in flexible ducts can also increase pressure drops up to a factor of nine. This resulted in a loss of ventilation rate and a significant increase in power and energy consumption by the HVAC system (Abushakra et al. 2003). Building airtightness is another area where the quality of the construction and the design of the building are as important as the materials in determining the desirable airtightness of the building envelope (Sherman and Chan 2006).

Energy and Costs

Ventilation requires energy not only to move the air but also to condition the supply air. Also, costs are involved for purchasing, designing, and installing the equipment. Energy use for ventilation and infiltration is significant and can account for one-third to one-half of the total space-conditioning energy (Sherman and Matson 1993). Building energy uses account for approximately 40% of total primary energy use in developed countries. Of this, the residential sector uses 60% to 70% for space conditioning (Orme 1998). Practical measures can be taken to conserve energy while still providing healthy ventilation rates. These include avoiding unnecessary air changes (due to leaky buildings), using good control strategies (not opening windows during periods of heating and cooling), and optimising fan and equipment efficiencies. Orme (2001) has indicated that energy losses from air change are as important as conduction and equipment losses. Mechanical ventilation systems reduce energy costs related to conditioning supply air if the building envelope is tight and infiltration is limited. Sherman and Matson (1993) estimated that 2.1 EJ per year could be saved by tightening the existing US housing stock. Most of the US housing stock uses infiltration as the ventilation system. The average ventilation rate has been estimated at more than 1 ach, with an estimated energy load of 4 EJ annually. If the existing housing stock were tightened and a continuous mechanical ventilation system were installed to provide an national average air change rate of 0.52 ach, the researchers estimated the energy load to be 1.8 EJ with a cost savings of $2.4 billion (Sherman and Matson 1997). Energy consumption can be reduced 9% to 21% by installing a mechanical ventilation system with heat recovery (Hekmat et al. 1986).

Besides energy costs, ventilation systems involve the extra cost of purchasing and installing the equipment. The costs for various supply and exhaust mechanical systems. The ventilation systems are run continuously, but a cost estimate is made for running the central fan for mixing purposes (this may be an option for some houses).

The results show that a single-point exhaust system is the least expensive to purchase and install, with an estimated total cost of $72. This is supported by Wray et. al. (2000), who also found a mechanical exhaust system to be the least expensive to operate. While a four-point energy recovery ventilation system would be the most expensive to purchase and install ($1772), the benefits of improved air distribution, filtration opportunities, and energy savings may outweigh the initial costs. As expected, retrofitting an existing house is more expensive than new construction, and multi-point distribution systems were more expensive than a single-point system. If the house has an existing central fan system, then it need not be cost-prohibitive to integrate a supply ventilation system with a single-point exhaust.

Even though cost estimates are available, they are not necessarily sufficient to enable optimal selection of the ventilation system. Individual users may place high values on criteria that were not considered or heavily weighted. To evaluate such a multi-objective system requires exotic optimisation approaches. For example, Roberson et al. (1998) developed a unique optimisation for overall cost-effectiveness of a ventilation system (which included considerations for installation costs, operating costs, distribution effectiveness, and the potential for depressurisation and for condensation). A multi-point supply system was found to be the best system overall. In cold climates, the group recommended a balanced system (multi-point supply with single-point exhaust) to prevent moisture problems in the building walls. In most of these cases, however, a simple continuous exhaust system would have proven to be more cost-effective if the only objective was to meet Standard 62.2-2004.

Climate can have a large impact on energy use. In hot, humid climates, dehumidification is necessary in houses with controlled ventilation systems. According to Rudd et al. (2003), mechanical ventilation with a separate dehumidification system provided the best overall value, including humidity control, installation costs, and operating costs. Some key factors contributing to the energy savings were locating the ducts inside the conditioned space, using insulation, and installing high-performance windows.

Controls

A variety of control systems from the simple to the complex are available to adjust the ventilation rate to achieve comfort and energy savings. Available systems include timers, occupant sensors, C[O.sub.2] sensors, and outside temperature or humidity sensors. The least reliable system is relying on the occupant to open and/or close windows. The occupant will respond to odors, drafts, noise, or a need for privacy rather than the need for a certain ventilation rate (Liddament 2001). The area of residential ventilation controls will continue to grow as users wish to take advantage of intermittent ventilation options, to have pollutant or weather-sensitive mechanical systems, etc.

Distribution Systems

A distribution system provides uniform ventilation and is an important component of all ventilation systems. In general, central exhaust systems and natural and passive ventilation systems do not distribute the fresh air as well as a multi-point supply system or a mechanical system that uses the ductwork of an existing HVAC system (Rudd and Lstiburek 2000). These systems allow the supply air to enter the building envelope in a rather uncontrolled manner and inevitably some rooms do not receive enough air, while others are overventilated.

The distribution system is an integral part of many mechanical ventilation systems and can have a significant effect on the ventilation rate and efficiency of a building. Leaky ducts are a source of energy loss, ventilation rate loss, and, in the case of return ducts, a source of indoor pollution (Delmotte 2003). In particular, the location of the ductwork is important. Modera (1993) has shown an energy loss of 30% to 40% when ductwork is installed in unconditioned spaces. He also showed through field testing and modelling that leakage through the average duct system was 37% higher than infiltration through the building envelope. Houses with leaky ductwork and air handlers located outside the conditioned space are at risk for increased infiltration rates, especially in hot, humid climates. This has large impacts on the actual ventilation rate found in the average house. The ventilation rate in many houses may not meet ASHRAE standards even though the equipment was designed to provide adequate ventilation, since leaky ductwork can prevent effective distribution of the supply air.

One strategy to save conditioning energy is to close the registers or grilles in rooms that are not being used. This strategy can increase the pressure in the entire duct system and increases the leakage rate in the ducts. A recent study found that the energy saved due to conditioning the air was only partially offset by increased duct system losses (Walker 2003).

Indoor Air Quality

Exposure to indoor pollutants can pose a serious health risk, especially for sensitive populations such as the young, asthmatic, or elderly (Sherman and Hodgson 2004; Seppanen and Fisk 2004). Indoor pollution originates from both indoor and outdoor sources and may be in the form of suspended particulates, volatile organic chemicals (VOCs), human bio-effluents, and microbiological contaminants (Seppanen 2006). Occupant activities such as cooking, bathing, smoking, vacuuming, using cleaning products, painting, as well as chemical emissions from building materials, electrical equipment, and appliances are all examples of indoor sources. Outdoor sources primarily result from vehicle exhaust but also agricultural activities, construction, manufacturing activities, ground sources (radon), and allergens (Levin 2004). The most effective method for controlling pollutants is by reducing or eliminating the source of the emission, but this is not always possible for some pollutants (Sherman and Matson 2003; Levin 2004). A number of strategies exist for improving indoor air quality, including increasing ventilation rates to dilute the pollutant, filtration to remove particulates, or air cleaning to capture VOCs, or a combination of all three strategies. Proper maintenance and operation of the ventilation system, appropriate building design to limit sources of pollution, avoiding excessive depressurisation, providing local ventilation at sources that produce pollution (combustion appliances), and moisture control are all important strategies in controlling indoor air quality (Hadlich and Grimsrud 1999).

Dilution Ventilation. Appropriate whole-house ventilation can dilute the level of indoor pollutants with fresh outdoor air (assuming the outdoor air in not more contaminated than the indoor air). Diluting pollutants with more fresh air has historically been the function of ventilation; however, it is not a pollutant-specific strategy and not all pollutants can be treated the same way. Almost all of the ventilation technologies described previously can provide the necessary ventilation rates for effective dilution. For natural ventilation and/or passive systems there is some inherent lack of control of ventilation rates that may result in times when indoor pollution is high. Although these systems may provide an annual average acceptable ventilation rate, they cannot effectively deal with peak periods of pollution (Sherman and Wilson 1986). On the other hand, all mechanical systems offer high levels of ventilation rate control so that indoor pollutants can always be diluted. Plus, many mechanical systems also include local fans in areas where production of pollutants is high, such as bathrooms and kitchens, in order to minimise the spread of pollutants into other parts of the house. Also, mechanical systems can provide the higher ventilation rates that are required to dilute VOCs, such as formaldehyde, than those needed to control human bio-effluents, such as C[O.sub.2]. Filtration. Sherman and Matson (2003) have shown that dilution ventilation is not always effective at reducing particle concentrations. Effective filtration can reduce the concentration of particulates that cannot be reduced at the source; this can also reduce the need for ventilation dilution. There are several methods that can be used to reduce particulate levels, including filtration, electrostatic precipitators, and simply by the deposition that occurs in the HVAC system.

Thatcher and Layton (1995) showed that the shell of the building offers little if any filtration of total particles and that indoor particle concentrations are significantly impacted by the activity level of the residents in the house. Even light activity, such as walking, can significantly increase the suspended particulate concentration for supermicron particles. Wallace et al. (2004) showed that the use of a central fan in a forced-air system alone could reduce the whole-house particle concentration (P[M.sub.2.5]) by 14%, and that installation of an in-duct mechanical filter could reduce the levels of particles by 23%. An electrostatic precipitator can reduce particles, especially fine particles, by 51%, but these are more expensive than mechanical filters and require routine maintenance to remain effective.

Filtration is most commonly used in mechanically ventilated buildings with supply systems and can be used to filter recirculated air or to filter the incoming supply air. Particle filters are rated by the ASHRAE Minimum Efficiency Reporting Value (MERV) scale (ASHRAE 1999). Typical furnace filters are rated at MERV 4 or lower and are not effective at removing respirable particles, but they can remove large pollens and visible dust particles. MERV filters rated 6 to 8 can remove smaller particles in the range of 10 [micro]m (P[M.sub.10]) and filters with a MERV rating of 9h[micro] to 13 can remove fine respirable 2.5 [micro]m (P[M.sub.2.5]) particles. Currently Standard 62.2-2004 recommends using a MERV 6 filter to protect the HVAC system from accumulating particles and becoming itself a source of indoor pollutants.

Since residential HVAC systems operate cyclically, filters used as part of the HVAC system perform better when the fraction run-time is high. Fugler and Bowser (2002) showed that high-efficiency furnace filters have a minimal effect on indoor particulate (P[M.sub.10]) levels when the occupants are active, but during low activity times (sleeping), P[M.sub.10] could be reduced 70%.

Filtration performance is selective; it often has poorer efficiencies for the finest of particle sizes and will fail unless care is taken in the installation and maintenance of the system (Liddament 2001). If the building envelope is tight and the filtration system is maintained, there is a potential to reduce both the indoor particulate levels and the ingress of outdoor particulates into the indoor environment. According to Sherman and Matson (2003), a MERV 11 filter installed in a supply ventilation system can reduce cat and dust mite allergens 30% to 40%. They recommend installing a MERV 9 to 12 filter and reducing duct leaks, preventing filter bypass, reducing uncontrolled infiltration, and running the fan continuously in order to maximise the filtration efficiency. This comes at a high energy cost.

Outdoor pollution presents a serious limitation for naturally (or passively) ventilated buildings, especially in urban areas. Researchers have shown that outdoor particles penetrate fully (almost 100%) into the indoor environment of houses with very leaky building envelopes and/or open windows that do not provide much opportunity for interaction between the airstream and the envelope (Thatcher and Layton 1995; Thatcher et al. 2001; Partti-Pellinen et al. 2000). In a comparison of mechanical ventilation systems with and without filtration, however, the Canadian Mortgage and Housing Corporation (CMHC 2003) found that unfiltered exhaust systems provided some protection from outdoor particles when compared to unfiltered supply or balanced ventilation systems. These provided no protection from the ingress of outdoor particles. Emmerich and Nabinger (2001) found penetration factors of 60% to 80% in test houses.

These studies suggest that the building envelope offers some protection from pollen, allergens, and diesel particles. Ventilation systems that move the supply air through the building envelope (such as natural infiltration, passive systems, and mechanical exhaust systems) can provide some minimal filtration from these types of outdoor particles. The CMHC found that the best protection from outdoor particles was provided by a ventilation system that positively pressurised the house and used a high-efficiency particle filter (HEPA), which can be expensive. In this case, a HEPA filter

Evans, Chuck. Jewelry: *Contemporary Design and Technique*, Davis Publications, Worcester, MA, 1983.

Flynn, M.J. : *Computer Architecture: Pipelined and Parallel Processor Design*, Narosa, Delhi, 2011.

Francois Cardarelli: *Scientific Unit Conversion: A Practical Guide to Metrication*, Springer, M.J. Shields. London, 1997.

Gostelow, Mary: *The Fan*, London, 1976

Groover, Mikell P.: *Fundamentals of Modern Manufacturing*, John Wiley & Sons, INC, New York, 2002.

Herman, Stephen L.: *Electric Motor Control*, Cengage Learning, Delmar, 2009.

Hobson, A.: *Fans: Concepts and Connections*. Englewood Cliffs, NJ: Prentice-Hall, 1995.

James Brown: *Advanced Machining Technology Handbook,* McGraw-Hill, New York, 1998.

Joseph A. Brink Jr.: *The Chemical Process Industries*, McGraw Hill, New York, 1977,

Kern, D. Q.: *Process Heat Transfer*, McGraw-Hill, New York, 1950.

Liepmann, H., and A. Roshko: *Elements of Gas Dynamics*. John Wiley Publishers, 1957.

MacCullough, G.H.: *Elements of Strength of Materials,* Van Nostrand Co., New York, 1957.

Massart, D. L.: *Chemometrics: A Textbook*, Amsterdam, Elsevier, 1988.

Mayor, Susan: *Collecting Fans*, London, 1980

Pradeep P. Phule: *The Science & Engineering of Materials*, Thomson-Engineering, NY, 2005.

Shigley, J.E.: *Mechanical Engineering Design,* McGraw-Hill, Inc., New York, 1972.

Singer, C.: *A History of Technology,* The Clarendon Press, Oxford, 1967.

Singh K & Soler A: *Mechanical Design of Heat Exchangers*, Arcturus, London, 1984

Thompson, P. A.: *Compressible Fluid Dynamics*. Maple Press Company, 1984.

Van Vlack, L. H.: *Elements of Materials Science,* Addison-Wesley Publishing Co., Reading, Mass., 1967.

Wayne, Beaty: *Standard Handbook for Electrical Engineers,* McGraw-Hill, New York, 1978.

Yahya, S. M. : *Turbines, Compressors and Fans*, Tata McGraw Hill, 2005.

Index

F

H

I

L

M

P

R

S

T

W

□□□